The Quantum Society

How Quantum Computing Will Reshape Everything

Laura Lee

Lauxon Publishing, LLC
Info@Lauxonpublishing.com
www.Lauxonpublishing.com

CONTENTS:

Chronicle the milestones of quantum supremacy—what it is, who achieved it, and what it means for the future of technology and society.

Chapter 5: Quantum Algorithms: Beyond the Binary

Exploring the quantum algorithms (like Shor's and Grover's algorithms) that will unlock new possibilities in problem-solving across industries, from cryptography to optimization.

Chapter 6: Quantum Cryptography: The End of Secrets?

How quantum computing will disrupt encryption. A look into quantum cryptography, including quantum key distribution (QKD) and the vulnerability of current encryption methods.

Chapter 7: A New Age of Cybersecurity

The role of quantum computing in reshaping cybersecurity infrastructure and potential solutions to protect data in the quantum era.

Chapter 8: Quantum and AI: A New Partnership

Exploring how quantum computing will fuel advancements in artificial intelligence and machine learning, making systems smarter, faster, and more capable of processing enormous data sets.

Chapter 9: The Quantum Internet: Reimagining Connectivity

The concept of a quantum internet and its potential to revolutionize communication with unhackable networks, instant information transfer, and secure data sharing.

Chapter 10: Quantum Medicine: The Future of Health

How quantum computing will transform the healthcare industry—faster drug discovery, personalized medicine, better diagnosis, and quantum simulations of biological systems.

Chapter 11: The Future of Climate Science

Examining how quantum computing will tackle climate change, from optimizing renewable energy systems to modeling complex climate systems in ways that classical computers cannot.

Chapter 12: Transforming Material Science and Chemistry

Quantum simulations will enable breakthroughs in materials science and chemistry, leading to the discovery of new materials and compounds with revolutionary properties.

Chapter 13: Quantum Finance: Unleashing New Markets

Quantum's impact on financial systems, from optimizing portfolios to predicting market behaviors and transforming risk management models.

Chapter 14: Supercharging Supply Chains

Quantum's potential to optimize complex supply chains, reduce inefficiencies, and respond to real-time global challenges in logistics and transportation.

Chapter 15: Quantum in Agriculture: Feeding a Growing World

Exploring how quantum computing can address global agricultural challenges, from optimizing crop yields to managing food supply chains more effectively.

Chapter 16: Quantum Biology: Understanding Life Itself

Quantum computing's role in solving some of the most complex problems in biology, from understanding photosynthesis to decoding protein folding and cellular behavior.

Chapter 17: Quantum Simulations: A Window into the Universe

Using quantum computers to simulate the behavior of particles at the quantum level, helping us understand phenomena in physics, chemistry, and cosmology.

Chapter 18: Quantum Energy: Redefining Power Systems

How quantum technologies could revolutionize energy systems, from quantum batteries to optimizing power grids and unlocking new renewable energy sources.

Chapter 19: The Ethics of Quantum Computing

What are the ethical considerations of quantum computing? This chapter will explore privacy, inequality, and the potential misuse of quantum technologies.

Chapter 20: Democratizing Quantum: Ensuring Global Access

Addressing the risks of a quantum divide, where access to quantum computing may be limited to the wealthy or privileged nations, and proposing solutions for global inclusion.

Chapter 21: Quantum Governance: Policymaking in the Quantum Age

How governments and international bodies will need to regulate quantum technologies to prevent misuse while encouraging innovation.

Chapter 22: Quantum Warfare: The New Arms Race

Quantum computing as a tool in modern warfare—defense strategies, potential offensive uses, and the global race for quantum dominance.

Chapter 23: Quantum Diplomacy: Global Cooperation

The need for international collaboration on quantum research, quantum cybersecurity, and ethical guidelines. The role of organizations like the UN and NATO.

Chapter 24: Quantum Computing in Space Exploration

How quantum computing will accelerate space exploration, from optimizing spacecraft design to enhancing communication systems across the cosmos.

Chapter 25: Redefining the Economy in the Quantum Era

The transformative impact of quantum computing on global economics, from labor markets to productivity and the future of innovation-driven economies.

Chapter 26: The Quantum Workforce: Jobs of the Future

How quantum computing will reshape the job market, the skills required, and the educational changes necessary to prepare future generations.

Chapter 27: Quantum Education: Redefining Learning

Exploring the need for new curricula and teaching methods in a world where quantum literacy is as important as traditional STEM skills.

Chapter 28: Quantum Computing in Entertainment

How quantum computing will change content creation, from more realistic simulations in gaming and movies to generating new forms of interactive storytelling.

Chapter 29: The Arts in the Quantum Age

How quantum computing might influence artistic expression and creativity, potentially generating new forms of digital art and immersive experiences.

Chapter 30: Quantum in Sports and Fitness

The impact of quantum computing on sports science—enhancing athlete performance through better data analysis and optimizing fitness training programs.

Chapter 31: Redefining Relationships: Quantum Psychology

Examining the potential intersection of quantum computing with neuroscience and psychology to deepen our understanding of the human mind, cognition, and behavior.

Chapter 32: Quantum Societies: How Communities Will Evolve

How quantum computing will influence social structures, global interconnectedness, and possibly the evolution of human societies as we know them.

Chapter 33: The Quantum City: Urban Planning and Smart Cities

Exploring how quantum technologies will optimize cities, from transportation systems to resource management and energy efficiency in smart cities.

Chapter 34: Quantum Governance in Smart Cities

The role of quantum in enabling new forms of governance in smart cities, using quantum computers to handle vast amounts of data for decision-making and crisis management.

Chapter 35: Quantum Communication and the Media

How quantum technology will revolutionize journalism, media, and the way information is disseminated and consumed by society.

Chapter 36: The Quantum Singularity: The Limits of Human Knowledge

Discussing the philosophical implications of quantum computing's potential to solve problems beyond human understanding and its impact on knowledge itself.

Chapter 37: The Quantum Renaissance: A New Age of Discovery

How quantum technologies will usher in a new scientific renaissance, revolutionizing fields from physics to biology, and inspiring future generations of innovators.

Chapter 38: Preparing for the Quantum Future

Practical advice on preparing for a world transformed by quantum computing, from corporate strategies to educational shifts and personal growth.

Chapter 39: The Dark Side of Quantum Computing

The potential dangers of quantum computing—whether it's economic disruption, ethical misuse, or the empowerment of malicious actors.

Chapter 40: Quantum Ecosystems: Building the Infrastructure

The need for a quantum ecosystem to support the widespread use of quantum computers, including hardware, software, quantum clouds, and developer communities.

Chapter 41: The Role of Tech Giants and Startups

Examining the major players in quantum computing—big tech companies, startups, and research institutions—and how they are shaping the future.

Chapter 42: Quantum Computing and Human Augmentation

How quantum computing could intersect with biotechnology, enhancing human cognitive and physical abilities, and transforming what it means to be human.

Chapter 43: Quantum Religion: Spirituality in a Quantum World

Exploring how quantum theories and discoveries could influence spiritual beliefs, philosophical outlooks, and humanity's relationship with the unknown.

Chapter 44: Quantum Governance: The Global Rulebook

Proposing new governance models to ensure responsible development and use of quantum technologies, addressing issues like global security, privacy, and social equity.

Chapter 45: Quantum Leadership: Innovators of Tomorrow

Profiling the thought leaders, scientists, and visionaries driving the quantum computing revolution and their vision for the future.

Chapter 46: Quantum and the Environment: Solving Global Crises

How quantum computing could help address critical environmental issues, from managing natural resources to fighting pollution and environmental degradation.

Chapter 47: Quantum Collaboration: Building a Better World

The need for global cooperation and interdisciplinary collaboration in advancing quantum technologies for the collective good.

Chapter 48: The Quantum Society: A New Social Contract

Envisioning how the social contract between individuals, governments, and corporations will evolve in a world where quantum computing reshapes every aspect of life.

Chapter 49: Quantum Futures: Utopias and Dystopias

Speculating on possible futures—a utopian society enhanced by quantum technology versus dystopian outcomes where the technology exacerbates existing social divides.

Chapter 50: Embracing the Quantum Age

A conclusion that synthesizes the book's main themes, encouraging readers to embrace the opportunities and challenges of the quantum era with hope and responsibility.

PROLOGUE:

The Dawn of the Quantum Age

In every era, there comes a breakthrough that fundamentally alters the course of human history. Fire gave us warmth, the wheel gave us mobility, electricity illuminated our nights, and the internet connected the globe in ways once unimaginable. Today, we stand on the precipice of another such revolution, a leap so profound it promises to transform not just technology, but the very fabric of society itself.

Welcome to the Quantum Age.

For decades, scientists, mathematicians, and engineers have toiled to uncover the mysteries of the quantum world—a realm where particles behave in strange, counterintuitive ways, seemingly defying the laws of classical physics. It is a place where particles can be in two places at once, where information can travel instantaneously, and where the rules of computation as we know them are rewritten.

Quantum computing, a concept once confined to the pages of theoretical physics and the minds of dreamers, is now poised to move from research labs into the real world. Unlike the computers of today, which process information using bits of binary code—ones and zeros—quantum computers harness the power of qubits, quantum bits that can exist in multiple states simultaneously. This ability unlocks a level of computational power so vast that it is difficult to fully grasp.

Imagine a machine that could solve problems in seconds that would take today's most advanced supercomputers millions of years. A world where new medicines are discovered not through trial and error, but by simulating biological processes at the molecular level in an instant. A future where financial markets, global supply chains, and energy grids are optimized perfectly, anticipating and solving problems before they even arise. A time when encryption and privacy, as we know them, are forever changed—broken down and rebuilt by the same technology.

Yet, the implications of quantum computing go far beyond technical marvels. This technology will ripple across every aspect of our lives, reshaping industries, challenging governments, and even altering the way we think about ethics, privacy, and equality. It promises to cure some of humanity's greatest ills—while also presenting dangers that could amplify our darkest tendencies.

Will quantum computing be the great equalizer, lifting humanity to new heights of collaboration and prosperity? Or will it become the domain of the few, increasing divides and sowing instability in a world already fraught with inequality?

As with all transformative technologies, the answers to these questions will depend not just on the capabilities of the machines, but on the choices we make as a society. The world that quantum computing creates will be the one we choose to build with it.

This book is not just a technical exploration of quantum computing, though it delves deeply into its intricacies. It is a journey into the future—one that examines how quantum technologies will impact everything from the way we heal, to how we govern, to the ethical decisions we will face in a rapidly changing world. It is about what happens when the impossible becomes possible, when the future arrives faster than we anticipated, and when society must adapt—or risk being left behind.

As you read these pages, consider this: quantum computing is no longer the future. It is the present, quietly unfolding. And soon, its effects will be felt everywhere—from the smallest corners of everyday life to the grandest challenges facing humanity.

The quantum revolution has begun. Are we ready?

CHAPTER 1:

The Quantum Leap

The world of technology is filled with incremental improvements—faster processors, sharper screens, more efficient algorithms—but every once in a while, a true leap occurs. A leap that redefines the rules of what is possible. Quantum computing represents one of those rare moments, a transformative leap that promises to reshape the technological landscape in ways we are only beginning to imagine.

For centuries, the story of human progress has been intertwined with our ability to compute, to process information, and to solve problems. The ancient abacus, mechanical calculators, and eventually the rise of digital computers have all pushed the boundaries of what we can accomplish. Classical computers—those that rely on transistors to manipulate bits of information, represented as ones and zeros—have fueled the digital age, enabling everything from the internet to artificial intelligence. But even as powerful as today's computers have become, they are still bound by certain limitations. Complex problems like modeling climate change, simulating the behavior of molecules for drug development, or breaking modern encryption codes remain beyond the reach of even the fastest supercomputers.

This is where quantum computing enters the story.

The Quantum Breakthrough

To understand why quantum computing is so revolutionary, we must first journey into the strange and counterintuitive world of quantum mechanics. Quantum mechanics is the branch of physics that describes the behavior of particles at the smallest scales—atoms, electrons, and photons. In this microscopic realm, particles can exist in multiple states at once, a phenomenon known as superposition. Additionally, particles can become "entangled," meaning their properties are interconnected, no matter how far apart they are. These strange behaviors seem to defy the laws of classical physics, and for a long time, they were considered more of a curiosity for theoretical physicists than something that could impact everyday life.

But in the 1980s, pioneering scientists like Richard Feynman and David Deutsch began to imagine a new kind of computer, one that didn't just operate on classical bits but harnessed the power of quantum mechanics. Instead of bits, which exist in one of two states (0 or 1), these machines would use quantum bits, or qubits, which can exist in multiple states simultaneously. This key difference unlocks an extraordinary computational potential.

A quantum computer can process vast amounts of information in parallel, thanks to the superposition of qubits. While a classical computer would need to perform each calculation one after the other, a quantum computer can explore many possibilities at once. It's like being able to read every page of a book simultaneously instead of flipping through one page at a time.

The Race to Quantum Supremacy

For decades, quantum computing remained largely theoretical. Building a machine that could harness and control the fragile quantum states of particles was a monumental challenge. Early attempts were beset by instability—qubits would "decohere," or lose their quantum properties, making the machines highly error-prone. However, steady progress was made in laboratories around the world, and by the early 21st century, the dream of a functional quantum computer was no longer science fiction.

In 2019, Google made headlines when it announced that it had achieved "quantum supremacy." This milestone, though somewhat controversial, marked the moment when a quantum computer successfully performed a calculation that would have taken the world's most powerful classical computer thousands of years to complete. Though the specific problem solved by Google's quantum processor wasn't particularly useful in a practical sense, it demonstrated the immense potential of quantum technology. It was proof that quantum computing had moved from the realm of theory into reality.

Since then, other companies like IBM, Microsoft, and startup leaders like Rigetti and IonQ have entered the race to build the first fully practical quantum computer. Quantum research is also booming in academia, with universities and research institutions across the globe contributing to breakthroughs in qubit stability, error correction, and algorithm development.

The Quantum Revolution: Why It Matters

At first glance, quantum computing might seem like just another advancement in the long line of computing innovations, but the reality is far more profound. The leap from classical to quantum computing is akin to the shift from steam engines to electricity, or from horses to the internal combustion engine. It will fundamentally reshape industries, economies, and even the structure of society itself.

One of the most immediate and significant impacts of quantum computing will be in the field of cryptography. Much of the encryption that protects modern digital communications, including online banking, government secrets, and personal data, relies on the difficulty of factoring large numbers. Classical computers struggle with this problem, making modern encryption relatively secure. However, quantum computers, with their ability to perform parallel calculations, could break these codes in a fraction of the time, rendering many of today's security protocols obsolete. This has led to a new field known as post-quantum cryptography, as governments and organizations scramble to develop new encryption methods resistant to quantum attacks.

But the potential applications of quantum computing extend far beyond breaking codes. In chemistry and material science, quantum computers could simulate molecular structures with unparalleled precision, unlocking new materials with extraordinary properties, from superconductors to more efficient solar cells. In healthcare, quantum simulations could lead to faster drug discovery by modeling complex biological processes at the molecular level, helping to cure diseases that have eluded modern medicine. Industries as varied as finance, logistics, and artificial intelligence stand to be revolutionized by the power of quantum computation.

Consider the problem of optimizing supply chains, a notoriously complex issue involving countless variables—demand forecasting, transportation routes, labor availability, and production capacities. Classical computers can only approximate solutions to such problems, but quantum computers could find optimal solutions in real-time, reducing costs, improving efficiency, and even lowering the environmental impact of global trade.

The Challenges Ahead

While the promise of quantum computing is immense, we are still in the early days of this technological revolution. Current quantum computers, known as noisy intermediate-scale quantum (NISQ) devices, are powerful but still limited by instability and error rates. Building large-scale, fault-tolerant quantum computers that can reliably perform useful tasks will take time and continued investment. The field faces a range of technical hurdles, from improving qubit coherence times to developing effective quantum error correction techniques.

Moreover, the social and ethical implications of quantum computing cannot be overlooked. Who will have access to this powerful technology? Will it widen the gap between rich and poor, or between nations? What kind of governance and international cooperation will be needed to ensure that quantum computing benefits humanity as a whole, rather than being used for harmful purposes?

The Quantum Society

As with any leap in technology, the rise of quantum computing will not occur in isolation. It will intersect with other emerging fields—artificial intelligence, biotechnology, renewable energy—and reshape the future of society in ways that are difficult to predict. The quantum revolution is not just a technical shift; it is a societal one. Just as the internet transformed how we communicate, work, and live, quantum computing will redefine the limits of human possibility.

CHAPTER 2:

What is Quantum Computing?

At the heart of every major technological leap lies a shift in how we process and manipulate information. Classical computers, the machines that power everything from your smartphone to the largest supercomputers in the world, are built on a system of binary code. This is a language of ones and zeros—bits—that form the foundation of modern computation. But in quantum computing, the rules change. Instead of bits, we deal with qubits. Instead of binary states, we encounter superposition. And instead of simple logic gates, we manipulate quantum systems using quantum gates. These concepts are the building blocks of quantum computing, and they defy many of the principles that have guided classical computing for over a century.

To truly appreciate how quantum computing is set to revolutionize technology, we need to understand its foundational concepts—

qubits, superposition, entanglement, and quantum gates—and why they offer a profound departure from the classical paradigm.

Qubits: The Heart of Quantum Computing

In classical computing, the bit is the smallest unit of information. It can exist in one of two states: 0 or 1. This simple binary system allows computers to perform complex calculations by processing long strings of bits in sequences. In quantum computing, however, the basic unit of information is the quantum bit, or qubit.

A qubit, like a classical bit, can represent a 0 or a 1—but with a twist. Thanks to the principle of superposition, a qubit can also exist in a state that is both 0 and 1 at the same time. This is not merely a vague or probabilistic state; rather, the qubit exists in a specific combination of both possibilities, governed by quantum mechanics. Imagine a spinning coin: while in motion, it's neither heads nor tails but exists in a superposition of both. Only when the coin stops spinning does it "collapse" into one specific outcome. Similarly, a qubit remains in superposition until it is measured, at which point it assumes a definite value, either 0 or 1.

This property of superposition allows quantum computers to process an extraordinary number of possible outcomes simultaneously. Where a classical bit can only be in one state at a time (0 or 1), a qubit can represent multiple states simultaneously. This parallelism is what gives quantum computers their immense power. A single qubit in superposition can process two states at once. Two qubits can process four states, three qubits can process eight, and so on, exponentially increasing the computational possibilities.

Superposition: Computing in Multiple Dimensions

Superposition is one of the most fascinating and counterintuitive principles of quantum mechanics. To understand how it works, let's consider a more practical example: a classical computer solving a maze. A classical machine would explore one path at a time, checking if it leads to the exit, and if not, returning to try another. A quantum computer, on the other hand, could explore all possible paths simultaneously due to superposition, finding the correct route exponentially faster.

This ability to represent and compute multiple states at once is what allows quantum computers to tackle problems that are infeasible for classical computers. It's the key to understanding why quantum algorithms, such as Shor's algorithm for factoring large numbers, are so much faster than their classical counterparts.

However, superposition alone is not enough. The power of quantum computing also relies on another strange and powerful phenomenon: entanglement.

Entanglement: The Quantum Connection

Entanglement is perhaps the most mysterious and counterintuitive concept in quantum mechanics. It refers to a special relationship between two or more qubits, where the state of one qubit is directly related to the state of another, no matter how far apart they are in space. When two qubits become entangled, they form a single, interconnected system. If you measure the state of one entangled qubit, you instantaneously know the state of the other—even if they are separated by vast distances. This behavior puzzled even

Albert Einstein, who famously referred to it as "spooky action at a distance."

Entanglement is crucial for quantum computing because it allows qubits to share information in ways that classical bits cannot. In a classical computer, each bit is independent of the others. In a quantum computer, however, entangled qubits can work together in a kind of quantum harmony, solving problems cooperatively. The collective power of entangled qubits enables quantum computers to solve certain problems exponentially faster than any classical computer ever could.

To illustrate, let's return to our maze example. With entangled qubits, a quantum computer doesn't just explore all possible paths in parallel; the qubits are also "aware" of each other's results. This interdependence allows the quantum computer to arrive at the solution far more efficiently than classical computation could ever hope to achieve.

Quantum Gates: Manipulating Qubits

In classical computing, bits are manipulated using logic gates like AND, OR, and NOT gates, which perform operations on the bits to produce new outputs. Quantum computers have their own version of these gates, called quantum gates. These gates operate on qubits and leverage their quantum properties to perform computations that would be impossible for classical logic gates.

Quantum gates manipulate qubits by changing their probability amplitudes—the likelihood that a qubit will be in a certain state when measured. Some of the most important quantum gates include:

Hadamard Gate (H): This gate puts a qubit into a superposition state, creating an equal probability of it being in state 0 or 1. The Hadamard gate is often used to initialize quantum algorithms by creating superpositions that allow for parallel computation.

Pauli-X Gate (X): Analogous to a classical NOT gate, the Pauli-X gate flips the state of a qubit. If a qubit is in state 0, the X gate changes it to 1, and vice versa.

CNOT Gate (Controlled NOT): This is an entangling gate that operates on two qubits. It flips the state of the second qubit (the target qubit) if the first qubit (the control qubit) is in state 1. The CNOT gate is a building block for creating entangled qubits.

Phase Gate (S, T): These gates modify the phase of a qubit without changing its superposition probabilities. Phases are crucial in quantum algorithms as they determine how qubit states interfere with one another to produce the correct results.

Quantum gates are reversible, meaning that the operation can be undone to retrieve the original state of the qubits. This property is essential for quantum error correction and for preserving the delicate quantum states required for computation.

Quantum vs Classical: A New Paradigm

At this point, it's important to ask: What makes quantum computing so different from classical computing? After all, both types of machines are just manipulating information, right? The answer lies in how they approach problem-solving.

Parallelism vs Serialism: Classical computers are inherently serial machines. They process one piece of information at a time, regardless of how fast they do it. Quantum computers, however, are parallel by nature. Because of superposition, quantum computers can explore many possible solutions simultaneously, dramatically reducing the time required to solve certain problems.

Entanglement: Classical computers rely on independent bits. Each bit exists in isolation, performing its own function without knowledge of other bits. Quantum computers, through entanglement, allow qubits to communicate instantaneously and work together as part of a larger system, enhancing computational efficiency and accuracy.

Exponential vs Linear Scaling: Classical computers struggle with scaling. As problems grow more complex, the time required to solve them increases dramatically, often exponentially. Quantum computers, on the other hand, can scale exponentially with the number of qubits. This exponential scaling is what gives quantum computers the potential to outperform even the fastest classical machines on certain tasks.

Error Correction: Classical computers rely on redundancy and error correction codes to ensure reliability. Quantum computers, dealing with the delicate nature of qubits, require a completely different approach. Quantum error correction is an area of active research, as it's crucial for developing practical, large-scale quantum computers.

Challenges and Limitations

While the potential of quantum computing is enormous, it's important to recognize that we are still in the early stages of this technology. Current quantum computers are known as noisy intermediate-scale quantum (NISQ) devices, meaning they are small and prone to errors due to noise and decoherence. Qubits are incredibly sensitive to their environment, and even the slightest disturbance can cause them to lose their quantum properties, a process known as decoherence.

Building a large-scale, fault-tolerant quantum computer will require overcoming significant technical hurdles, including improving qubit coherence times, developing more efficient quantum gates, and perfecting quantum error correction techniques.

The Quantum Frontier

Despite these challenges, the quantum leap is well underway. As research advances and companies race to develop practical quantum machines, we are likely to see groundbreaking applications emerge across industries. From cryptography and drug discovery to artificial intelligence and climate modeling, quantum computing promises to revolutionize fields that rely on complex problem-solving.

In the coming chapters, we will explore how quantum computing will impact various sectors, from finance to healthcare, and consider the ethical and societal implications of a technology that promises to reshape the world.

Quantum computing is not just a new tool—it is a new way of thinking about computation. It is a leap into a realm where the rules of classical logic no longer apply, and where the boundaries of possibility are being redrawn before our eyes. The age of quantum is upon us, and it is set to change everything.

CHAPTER 3:

Breaking Boundaries: Quantum vs Classical Computing

For decades, classical computing has driven technological advancement, pushing the boundaries of what we thought possible. From the dawn of the personal computer to the age of supercomputers and cloud computing, classical machines have enabled unprecedented growth in information processing, artificial intelligence, communication, and industry. But as we enter the 21st century, the limits of classical computing are becoming evident. Some of the most complex problems—climate modeling, drug discovery, and advanced cryptography—remain stubbornly out of reach for even the most powerful classical machines.

Quantum computing represents the next great leap in the evolution of computational technology. It doesn't just offer more power; it fundamentally changes how computation is performed. To appreciate the significance of this leap, we must examine how classical and quantum computers differ in their approach, their potential, and their limitations.

Classical Computing: The Binary Foundation

Classical computers are based on a binary system of computation. Information is encoded in bits, which can exist in one of two states: 0 or 1. These bits are manipulated through logic gates (AND, OR, NOT) to perform calculations. The simplicity of binary systems has allowed engineers to build increasingly complex processors capable of executing trillions of operations per second. Classical computers have grown exponentially in speed and efficiency over the years, following Moore's Law—the observation that the number of transistors on a microchip doubles approximately every two years, leading to a corresponding increase in computational power.

However, classical computers are constrained by their reliance on bits. A bit can only represent a single piece of information at any given time: it's either 0 or 1. Complex problems, such as simulating molecular behavior or optimizing large systems, require analyzing a vast number of possibilities. In classical computing, this means performing calculations one by one, which quickly becomes impractical as the number of variables increases.

Classical computers also face physical limitations. As transistors get smaller and smaller, we are approaching the point where quantum effects begin to interfere with the behavior of transistors themselves. At the atomic level, the laws of classical physics break down, and quantum mechanics takes over. Thus, classical computers are not only limited in terms of solving certain types of problems—they may also be approaching the end of their ability to continue scaling through hardware improvements.

Quantum Computing: A New Paradigm

Quantum computing, on the other hand, operates under entirely different principles. It leverages the unique properties of quantum mechanics to process information in ways that are fundamentally impossible for classical computers. At the heart of this difference is the qubit, the quantum equivalent of a classical bit.

As we explored in the previous chapter, qubits can exist in a state of superposition, meaning they can represent both 0 and 1 simultaneously. This allows quantum computers to process many possibilities at once, providing massive parallelism. Additionally, qubits can become entangled, meaning the state of one qubit is directly related to the state of another, regardless of distance. This interconnectedness allows quantum computers to perform computations in ways that classical systems simply cannot replicate.

Key Differences Between Classical and Quantum Computing

To understand why quantum computing is the next step in computational evolution, let's break down the major differences between classical and quantum computing.

1. Information Representation: Bits vs. Qubits

Classical Computing: Information is encoded in bits, which can be either 0 or 1. A classical computer processes bits one at a time, requiring sequential steps to solve complex problems. Even the most advanced supercomputers are constrained by the need to perform operations in a step-by-step manner.

Quantum Computing: Quantum computers use qubits, which can exist in multiple states simultaneously due to superposition. This allows a quantum computer to perform many calculations in parallel. For example, while a classical computer with 2 bits can represent only one of four possible states at a time (00, 01, 10, or 11), a quantum computer with 2 qubits can represent all four states simultaneously.

2. Computational Power: Exponential vs Linear Growth

Classical Computing: Classical computers increase their power by adding more bits and faster processors, but this growth is linear. To double the computational power of a classical computer, you need to double the number of bits and the speed of its processors. For complex problems with many variables, the time required to compute a solution grows exponentially.

Quantum Computing: The power of a quantum computer grows exponentially with the number of qubits. If you add just one more qubit to a quantum system, you effectively double the amount of information that can be processed simultaneously. This exponential growth means that for certain types of problems, quantum computers could outperform classical machines by many orders of magnitude. For example, while a classical computer might take thousands or millions of years to solve certain problems, a quantum computer could do it in seconds.

3. Problem Solving Approach: Serial vs Parallel Processing

Classical Computing: Classical computers are serial processors, meaning they solve problems by evaluating each possible solution one at a time. This approach works well for problems with a manageable number of variables, but it becomes prohibitively slow for problems involving complex systems, such as simulating the behavior of molecules or finding optimal solutions in large datasets.

Quantum Computing: Quantum computers, thanks to superposition and entanglement, can evaluate multiple solutions in parallel. Instead of checking each possible solution one by one, a quantum computer can analyze all possible solutions at once, drastically reducing the time required to find the best result. This parallelism is especially useful for optimization problems, machine learning, and simulations of quantum systems, where the number of possible outcomes is enormous.

4. Error Tolerance and Stability

Classical Computing: Classical computers are highly stable, with built-in error correction mechanisms that ensure reliable operation. Errors in classical systems are rare and usually caused by hardware malfunctions or software bugs. These errors can be detected and corrected using established techniques.

Quantum Computing: Quantum computers are inherently more prone to errors because qubits are fragile and can easily lose their quantum state through interactions with the environment, a phenomenon known as decoherence. Quantum error correction is an active area of research, and while progress is being made, quantum computers are still far more susceptible to noise and instability than classical computers. This is one of the key challenges that must be overcome before quantum computers can be scaled up for practical, large-scale applications.

5. Applications: Solving the Unsolvable

The true potential of quantum computing lies in its ability to solve problems that are intractable for classical computers. These include:

Cryptography: One of the most famous examples of quantum computing's power is its potential to break modern encryption techniques. Many encryption systems, such as RSA, rely on the difficulty of factoring large numbers. Classical computers struggle with this task, but quantum algorithms like Shor's algorithm can factor large numbers exponentially faster, posing a threat to current encryption methods.

Drug Discovery and Materials Science: Quantum computers can simulate the behavior of molecules and atoms with high precision, something classical computers can only approximate. This ability could revolutionize drug discovery by allowing scientists to model complex biological interactions and develop new materials with novel properties.

Optimization: Many real-world problems, from supply chain logistics to financial modeling, involve optimizing a system with many interdependent variables. Classical computers often rely on approximation techniques to find good solutions, but quantum computers could find optimal solutions in far less time.

Artificial Intelligence: Quantum computing could accelerate machine learning algorithms by enabling faster training and more efficient pattern recognition, opening up new possibilities for AI development.

Classical Strengths and Quantum Challenges

Despite the immense potential of quantum computing, classical computers are still far superior in many ways. For most everyday tasks—browsing the internet, running applications, playing games—classical computers are more efficient and practical. Quantum computers are currently limited by several factors:

Error Rates: As mentioned, quantum computers are highly susceptible to errors due to noise and decoherence. While quantum error correction techniques are being developed, practical quantum computers still require significant improvements in stability and error tolerance.

Scalability: Building large-scale quantum computers with thousands or millions of qubits is a major challenge. Current quantum machines, known as noisy intermediate-scale quantum (NISQ) devices, have only a few dozen qubits and are prone to errors, limiting their usefulness for real-world applications.

Specialization: Quantum computers are not general-purpose machines like classical computers. They excel at solving specific types of problems, such as optimization and cryptography, but for many tasks, classical computers will remain the better choice.

The Path Forward: Quantum-Classical Hybrid Systems

The future of computing is likely to be a hybrid one, where classical and quantum computers work together to solve problems. Classical computers will continue to handle everyday tasks and manage general-purpose applications, while quantum computers will be used for specialized tasks that require their immense parallel processing power.

For example, quantum computers could be used to accelerate machine learning by solving complex optimization problems, while classical computers handle the data preprocessing and interpretation. Similarly, in cryptography, classical computers could manage secure communications, while quantum computers are used to develop and test new encryption methods resistant to quantum attacks.

Why Quantum is the Next Step

Quantum computing represents a fundamental shift in how we think about and perform computation. It's not just a faster or more powerful version of classical computing—it's a completely new paradigm that breaks through the boundaries of what classical machines can achieve. By leveraging the principles of superposition, entanglement, and quantum parallelism, quantum computers have the potential to solve problems that are impossible for classical systems.

CHAPTER 4:

The Road to Quantum Supremacy

The term "quantum supremacy" refers to the point at which a quantum computer can perform a calculation that is infeasible for any classical computer to achieve in a reasonable timeframe. This milestone marks a pivotal moment in the evolution of quantum computing, demonstrating that quantum systems can solve certain problems more efficiently than their classical counterparts. The journey to this landmark achievement has been marked by significant milestones, breakthroughs, and collaborations among researchers and technologists across the globe.

What is Quantum Supremacy?

Quantum supremacy is often described as the "first major milestone" in quantum computing, signifying the moment when a quantum computer performs a task that would be practically impossible for any classical computer, regardless of its processing power. It does not mean that quantum computers will replace classical ones; rather, it emphasizes that quantum computers are capable of solving specific problems much faster than classical machines.

The importance of quantum supremacy extends beyond academic interest. It has profound implications for fields such as cryptography, materials science, drug discovery, and artificial intelligence. Achieving quantum supremacy is a clear indication that we are entering a new era of computational capability—one that could redefine the boundaries of what is computable.

The Journey to Supremacy: Key Milestones

Theoretical Foundations (1980s-1990s)
The groundwork for quantum computing was laid in the 1980s and 1990s, when scientists and theorists began to explore the potential of quantum mechanics for computation. Key milestones during this period included:

Richard Feynman (1981): Feynman proposed the idea of a quantum computer, suggesting that classical computers would struggle to simulate quantum systems due to the complexity and exponential scaling of quantum states.

David Deutsch (1985): Deutsch formulated the concept of a universal quantum computer, laying the theoretical foundation for quantum computation and introducing the first quantum algorithm.

Peter Shor (1994): Shor developed a groundbreaking algorithm that could factor large numbers exponentially faster than classical algorithms, demonstrating the potential for quantum computers to outperform classical machines in specific tasks.

Lov Grover (1996): Grover's algorithm provided a quadratic speedup for unstructured search problems, showcasing additional advantages of quantum computing over classical systems.

First Quantum Algorithms and Experimental Realizations (2000s)

With the theoretical groundwork established, researchers turned their attention to building practical quantum systems. Key milestones included:

2001: The first experimental demonstration of Shor's algorithm was achieved by IBM and Stanford University, using a liquid-state NMR quantum computer to factor the number 15 into its prime components.

2007: The first implementation of Grover's algorithm was demonstrated on a quantum system, further validating the potential of quantum computing for real-world applications.

Building Better Qubits (2010s)

As the field advanced, researchers focused on developing more stable and scalable qubits. Significant milestones during this period included:

2012: The first demonstration of quantum error correction was achieved, a crucial step towards building fault-tolerant quantum computers capable of performing complex calculations.

2015: IBM launched its quantum computing project, introducing the IBM Quantum Experience—a cloud-based platform allowing researchers and developers to experiment with quantum algorithms on actual quantum hardware.

2017: Google announced the development of the 72-qubit Bristlecone processor, designed to test the limits of quantum supremacy.

The Quantum Supremacy Claim (2019)
The milestone that brought the concept of quantum supremacy into the spotlight occurred in October 2019 when Google announced that it had achieved quantum supremacy with its 53-qubit processor, Sycamore.

Sycamore's Experiment: The Sycamore processor performed a complex sampling task—specifically, generating random numbers from a quantum circuit—that took approximately 200 seconds to complete. Google claimed that the equivalent task would take the most powerful classical supercomputers, such as the Summit, around 10,000 years to complete.

Peer Review and Controversy: While Google's claim was celebrated within the quantum community, it also sparked debate. IBM disputed the result, arguing that classical systems could perform the same task in less than two and a half days, emphasizing the need for careful evaluation and replication of results in scientific research.

A New Era of Quantum Research (2020s and Beyond)
Following Google's announcement, the quantum computing landscape experienced rapid growth and innovation. Major technology companies, academic institutions, and startups accelerated their research efforts, driven by the promise of quantum computing. Key developments during this period included:

Research Collaborations: Numerous partnerships emerged between tech giants, universities, and government agencies aimed at advancing quantum technologies, driving research in quantum algorithms, hardware, and applications.

Investment and Commercialization: Substantial investments flowed into the quantum computing sector, with companies like IBM, Microsoft, Amazon, and Rigetti Computing launching ambitious quantum computing initiatives. Startups began to emerge, focusing on developing quantum software, hardware, and applications.

Quantum Software Development: The focus shifted towards creating software that can effectively harness the power of quantum systems. Platforms like IBM's Qiskit and Google's Cirq provide developers with the tools to design quantum algorithms and run them on quantum hardware.

Progress in Quantum Error Correction: Continued advancements in error correction and qubit coherence have paved the way for more robust quantum systems, moving closer to practical, large-scale quantum computers.

Implications of Quantum Supremacy

Achieving quantum supremacy is more than a technical milestone; it signals a shift in the landscape of technology and society. The implications of this achievement extend across various domains:

Cryptography: One of the most immediate concerns is the impact on cryptography. Quantum computers have the potential to break widely used encryption schemes, such as RSA and ECC, which secure online communications and transactions. As quantum computing becomes more prevalent, there is an urgent need for developing quantum-resistant encryption methods.

Drug Discovery and Materials Science: Quantum computers could revolutionize fields such as drug discovery, allowing researchers to model complex molecular interactions with unprecedented accuracy. This capability could lead to the development of new medications and materials, addressing pressing global challenges.

Artificial Intelligence and Machine Learning: Quantum computing can enhance AI and machine learning by accelerating data analysis and optimizing algorithms. This could lead to significant advancements in natural language processing, computer vision, and predictive analytics.

Optimization Problems: Industries reliant on optimization, such as logistics and finance, could benefit from quantum computing's ability to solve complex optimization problems much faster than classical systems, improving efficiency and reducing costs.

Education and Workforce Development: The rise of quantum computing will necessitate a new workforce skilled in quantum technologies. Educational institutions will need to adapt curricula to prepare students for careers in quantum computing, data science, and related fields.

The Road Ahead

While the achievement of quantum supremacy is a significant milestone, it is essential to recognize that it is only the beginning of a long journey toward practical quantum computing. The challenges of building scalable, fault-tolerant quantum systems and developing algorithms that can exploit their full potential remain.

As researchers and engineers continue to push the boundaries of quantum technology, we can expect a new wave of innovation that transforms industries, reshapes society, and expands our understanding of computation itself. The future of quantum computing is bright, but it will require collaboration, investment, and a commitment to addressing the ethical and societal implications of this powerful technology.

CHAPTER 5:

Quantum Algorithms: Beyond the Binary

As we embark on the journey into the world of quantum algorithms, it's essential to understand that these algorithms represent a profound shift in the way we think about computation. Unlike classical algorithms, which are designed for traditional binary systems, quantum algorithms leverage the unique properties of quantum mechanics—such as superposition, entanglement, and interference—to solve problems more efficiently. This chapter will explore some of the most significant quantum algorithms, including Shor's and Grover's algorithms, and their transformative potential across various industries.

The Quantum Advantage

Before delving into specific algorithms, it is vital to grasp the concept of quantum advantage. This term refers to the situation where a quantum algorithm can outperform the best-known classical algorithms for a particular problem. Quantum advantage arises from the fundamental principles of quantum mechanics that allow for parallelism, complex interference, and the manipulation of qubits in ways that classical bits cannot achieve.

In many instances, quantum algorithms can exponentially reduce the time complexity for solving specific problems. This capability can unlock new possibilities for industries ranging from cryptography to optimization, enabling us to tackle challenges that were previously deemed intractable.

1. Shor's Algorithm: A Breakthrough in Cryptography

One of the most celebrated quantum algorithms is Shor's algorithm, developed by mathematician Peter Shor in 1994. Shor's algorithm revolutionized the field of cryptography and demonstrated the potential power of quantum computing. Its significance lies in its ability to factor large integers exponentially faster than any known classical algorithm.

The Problem of Factorization: Many encryption schemes, particularly RSA (Rivest-Shamir-Adleman), rely on the difficulty of factorizing large integers into their prime components. Classical algorithms, such as the general number field sieve, become computationally impractical as the size of the integers increases. For example, factoring a 2048-bit number could take classical computers thousands of years.

Shor's Algorithm: Shor's algorithm utilizes quantum mechanics to achieve a polynomial-time solution for factoring. The algorithm consists of two main parts:

Reduction to Period Finding: Shor's algorithm reduces the problem of factoring to finding the period of a function related to modular arithmetic. This step is performed using quantum techniques to efficiently identify the periodicity of a function.

Classical Post-Processing: Once the period is found, classical methods are used to derive the factors of the original integer. The quantum part of the algorithm drastically reduces the time complexity to polynomial time, making it feasible to factor large numbers.

Implications for Cryptography: The implications of Shor's algorithm are profound. If large-scale quantum computers become a reality, many current encryption methods could become obsolete. This has led to a race for developing post-quantum cryptography—encryption techniques that are secure against quantum attacks. Researchers are actively exploring new algorithms and protocols that can withstand the threat posed by quantum computing.

2. Grover's Algorithm: Revolutionizing Search Problems

Another groundbreaking quantum algorithm is Grover's algorithm, introduced by Lov Grover in 1996. Unlike Shor's algorithm, which is tailored to specific problems in cryptography, Grover's algorithm is designed for unstructured search problems. It provides a quadratic speedup compared to classical search algorithms.

The Search Problem: In classical computing, if you want to search for a specific item in an unsorted database of Nentries, the best approach requires examining each entry one by one, leading to an average time complexity of O(N). Grover's algorithm can perform this search in O(√N) time, effectively halving the search time.

How Grover's Algorithm Works: Grover's algorithm employs a combination of quantum superposition and interference to locate the desired item efficiently. The main steps of the algorithm are:

Initialization: The algorithm begins by preparing a superposition of all possible states representing the database entries.

Oracle Query: An oracle—a quantum subroutine—marks the correct solution by flipping the sign of its amplitude. This step is critical, as it differentiates the correct item from the rest.

Amplitude Amplification: The algorithm amplifies the probability amplitude of the marked solution through interference, making it more likely to be observed when measuring the quantum state.

Measurement: Finally, the algorithm measures the state, revealing the correct solution with a high probability.

Applications of Grover's Algorithm: The quadratic speedup provided by Grover's algorithm has significant implications across various domains:

Cryptography: Grover's algorithm can effectively reduce the security of symmetric-key encryption schemes, such as AES (Advanced Encryption Standard). While it does not break encryption, it reduces the effective key length—making a 256-bit key effectively equivalent to a 128-bit key against a quantum adversary.

Database Search: Grover's algorithm can enhance search algorithms in fields such as data mining, artificial intelligence, and bioinformatics, enabling faster and more efficient searches through large datasets.

Optimization Problems: Many optimization problems can be framed as search problems, and Grover’s algorithm can be used to identify optimal solutions more quickly than classical methods.

3. Other Notable Quantum Algorithms

While Shor's and Grover's algorithms are among the most famous, several other quantum algorithms also hold promise for various applications:

Quantum Fourier Transform (QFT): This algorithm is a crucial component of Shor’s algorithm and has applications in signal processing and solving differential equations. The QFT can perform Fourier transforms exponentially faster than classical algorithms.

Variational Quantum Eigensolver (VQE): VQE is a hybrid quantum-classical algorithm designed to find the lowest eigenvalue of a Hamiltonian, making it particularly useful for simulating molecular systems and materials. It combines classical optimization techniques with quantum computations, allowing for practical applications in quantum chemistry.

Quantum Approximate Optimization Algorithm (QAOA): QAOA is another hybrid algorithm aimed at solving combinatorial optimization problems. It utilizes quantum superposition to explore potential solutions and classical optimization to refine them, offering a promising approach to tackling NP-hard problems.

4. The Future of Quantum Algorithms

The field of quantum algorithms is still in its infancy, and researchers continue to explore new possibilities and enhancements. As quantum hardware matures and error correction techniques improve, we can expect an increasing number of practical quantum algorithms to emerge.

Tailored Algorithms for Specific Industries: Industries such as finance, logistics, pharmaceuticals, and artificial intelligence will likely see the development of specialized quantum algorithms designed to address their unique challenges. Custom algorithms could optimize supply chains, accelerate drug discovery, or enhance financial modeling.

Cross-Disciplinary Collaborations: As quantum computing technology advances, collaborations between computer scientists, physicists, mathematicians, and industry experts will foster innovation. These interdisciplinary efforts will yield new algorithms and applications, driving the adoption of quantum computing across sectors.

Quantum Machine Learning: The intersection of quantum computing and machine learning is a burgeoning area of research. Quantum algorithms could significantly speed up training times and improve the accuracy of machine learning models, leading to breakthroughs in data analysis and pattern recognition.

Conclusion: Beyond the Binary

Quantum algorithms represent a new frontier in computation, unlocking possibilities that extend far beyond the binary constraints of classical computing. As we continue to explore the potential of quantum algorithms like Shor's and Grover's, we must also recognize the challenges that lie ahead.

The transition from theory to practical application will require collaboration, innovation, and a commitment to understanding the ethical implications of these powerful technologies. Quantum algorithms will not only reshape industries but will also redefine our understanding of computation itself.

CHAPTER 6:

Quantum Cryptography: The End of Secrets?

As quantum computing advances, it poses a significant threat to traditional encryption methods that safeguard our digital communications. The principles of quantum mechanics not only challenge existing cryptographic protocols but also pave the way for new forms of secure communication—most notably, quantum cryptography. In this chapter, we will explore how quantum computing will disrupt encryption, examine the vulnerabilities of current methods, and delve into the promises of quantum key distribution (QKD) as a potential solution.

The Cryptographic Landscape Before Quantum Computing

Before delving into quantum cryptography, it's crucial to understand the role encryption plays in securing our digital world. Encryption serves as the foundation of data privacy, protecting sensitive information in various applications—from online banking to secure communication. Traditional encryption methods, such as RSA and AES, rely on complex mathematical problems that are computationally difficult to solve using classical computers.

Public Key Cryptography: One of the most widely used forms of encryption is public key cryptography, exemplified by the RSA algorithm. This method uses a pair of keys—a public key for encryption and a private key for decryption. The security of RSA relies on the mathematical difficulty of factoring large numbers into their prime components.

Symmetric Key Cryptography: Symmetric encryption methods, like AES, use the same key for both encryption and decryption. The security of symmetric key algorithms depends on the length of the key, with longer keys providing greater security against brute-force attacks.

The Threat of Quantum Computing

As we discussed in Chapter 5, quantum computers have the potential to perform certain calculations exponentially faster than classical computers. This capability poses a direct threat to the foundational assumptions of current encryption methods.

Shor's Algorithm and Public Key Cryptography: Shor's algorithm can factor large integers in polynomial time, effectively breaking RSA encryption. If a large-scale quantum computer were to be built, it could compromise the security of any system relying on RSA or similar algorithms, allowing attackers to decrypt sensitive information and impersonate users.

Grover's Algorithm and Symmetric Key Cryptography: While Grover's algorithm does not break symmetric key encryption outright, it can reduce the effective key length. For example, a 256-bit key would offer equivalent security to a 128-bit key against a quantum adversary. This reduction significantly undermines the strength of symmetric encryption and necessitates longer key lengths to maintain security.

Quantum Cryptography: A New Paradigm

In response to the challenges posed by quantum computing, researchers have been exploring quantum cryptography as a promising alternative. Quantum cryptography leverages the principles of quantum mechanics to create secure communication channels that are theoretically immune to eavesdropping.

Quantum Key Distribution (QKD): The most well-known application of quantum cryptography is quantum key distribution (QKD). QKD allows two parties to generate a shared, secret key that can be used for secure communication. The security of QKD is based on the laws of quantum mechanics, specifically the behavior of quantum states.

The BB84 Protocol: Developed by Charles Bennett and Gilles Brassard in 1984, the BB84 protocol is the first and most widely studied QKD protocol. It utilizes polarized photons to transmit key information. The key steps of the BB84 protocol include:

Preparation of Quantum States: The sender (Alice) prepares a sequence of photons in one of four polarization states, representing binary values (0 or 1).

Transmission: Alice sends the polarized photons to the receiver (Bob).

Measurement: Bob measures the polarization of each photon, randomly choosing one of two bases for measurement.

Key Sifting: After transmission, Alice and Bob communicate over a classical channel to share their measurement bases. They discard any instances where their bases did not match, resulting in a shared secret key.

Eavesdropping Detection: If an eavesdropper (Eve) attempts to intercept the photons, the act of measuring the quantum states will disturb them due to the no-cloning theorem and the principles of superposition. This disturbance can be detected, allowing Alice and Bob to ensure the integrity of their shared key.

Security Advantages of QKD:

The No-Cloning Theorem: Quantum mechanics prohibits the cloning of unknown quantum states, meaning that an eavesdropper cannot create perfect copies of the quantum key being transmitted. This fundamental property enhances the security of QKD.
Disturbance Detection: The inherent properties of quantum states allow Alice and Bob to detect any unauthorized eavesdropping attempts. If Eve intercepts and measures the quantum states, the disturbance will reveal her presence.

The State of Quantum Cryptography Today

Although QKD offers a robust alternative to traditional encryption methods, it is still in its infancy. Several challenges need to be addressed before it can be widely adopted:

Technological Limitations: Current QKD systems are limited in range and speed. The transmission of quantum states is susceptible to loss and noise, which can degrade the quality of the key. Ongoing research aims to develop more robust QKD systems that can operate over longer distances and in various environments.

Integration with Existing Infrastructure: Implementing QKD alongside classical communication networks poses logistical challenges. Researchers are working on hybrid systems that can seamlessly integrate QKD with existing encryption protocols to enhance security.

Regulatory and Standardization Issues: As quantum cryptography emerges, regulatory bodies and industry leaders must establish standards and best practices to ensure interoperability and security across various platforms.

The Future of Quantum Cryptography

Despite the challenges, the potential of quantum cryptography is immense. As technology advances, we can expect several developments:

Increased Adoption: Industries that rely heavily on secure communication, such as finance, healthcare, and defense, are likely to lead the charge in adopting QKD and quantum cryptographic methods to protect sensitive data.

Post-Quantum Cryptography: As researchers work to develop quantum-resistant encryption methods, a hybrid approach that combines classical encryption with quantum cryptography could provide enhanced security. This may involve developing algorithms that can withstand quantum attacks while still being efficient for classical systems.

Quantum Networks: The establishment of quantum communication networks, often referred to as quantum internet, will facilitate the deployment of QKD on a larger scale. These networks would enable secure communication across vast distances, connecting quantum devices and enhancing global security infrastructure.

The End of Secrets?

The advent of quantum computing heralds a transformative era for cryptography, with the potential to disrupt existing encryption methods that underpin our digital security. While the vulnerabilities of classical encryption raise concerns, quantum cryptography offers a promising alternative through techniques like quantum key distribution.

As we navigate this new landscape, it is essential to strike a balance between the opportunities presented by quantum cryptography and the challenges that lie ahead. The development and implementation of quantum cryptographic systems will require collaboration across industries, research institutions, and governments to establish a secure digital future.

While quantum cryptography may not necessarily mark the end of secrets, it certainly represents a new paradigm in secure communication—one that could redefine our understanding of privacy and security in the quantum age. The journey towards a secure quantum future is just beginning, and as we continue to explore the implications of quantum technologies, we will uncover even greater possibilities for safeguarding our digital lives.

CHAPTER 7:

A New Age of Cybersecurity

As we step into the quantum era, cybersecurity faces unprecedented challenges. The rise of quantum computing not only threatens existing encryption methods but also calls for a fundamental rethinking of how we protect data, networks, and systems. In this chapter, we will explore how quantum computing will reshape cybersecurity infrastructure, the emerging threats, and the potential solutions to ensure security in the quantum age.

The Cybersecurity Landscape Today

In today's digital world, cybersecurity is vital for protecting personal data, financial systems, government infrastructure, and critical industries such as healthcare and energy. Modern cybersecurity defenses rely heavily on encryption, access controls, and secure communication protocols to protect sensitive information.

Encryption: At the core of cybersecurity is encryption, which scrambles data into unreadable formats unless decrypted with the correct key. This ensures data privacy and security, particularly during transmission over networks.

Firewalls and Intrusion Detection Systems (IDS): These systems help defend against unauthorized access, malware, and network-based attacks by monitoring and filtering traffic.

Authentication and Access Control: Multi-factor authentication, strong passwords, and role-based access are used to ensure only authorized individuals can access sensitive systems.

AI and Machine Learning: Increasingly, AI-powered tools are used to identify and respond to threats, learn from attack patterns, and improve defenses in real time.

However, as quantum computing becomes more powerful, many of these traditional methods face vulnerabilities that could be exploited by adversaries equipped with quantum computers.

The Quantum Threat to Cybersecurity

Quantum computers, with their ability to solve complex problems far faster than classical computers, bring new risks to current cybersecurity systems.

Breaking Encryption: As discussed in previous chapters, the most immediate quantum threat is to encryption. Shor's algorithm can factor large numbers exponentially faster than any classical algorithm, meaning that current public-key encryption methods like RSA and ECC (Elliptic Curve Cryptography) could be broken once a sufficiently powerful quantum computer is built.

RSA Vulnerability: The RSA encryption algorithm relies on the difficulty of factoring large numbers, a problem that quantum computers could solve in polynomial time. This could allow attackers to easily decrypt sensitive communications, leading to data breaches and exposure of personal, financial, and governmental secrets.

ECDSA Vulnerability: ECC-based methods like the Elliptic Curve Digital Signature Algorithm (ECDSA), which are widely used for secure transactions (e.g., in blockchain technologies), are similarly vulnerable to quantum attacks.

Symmetric Encryption Weakening: While Grover's algorithm does not completely break symmetric encryption like AES, it can reduce its security by half. For example, AES-256 would offer the equivalent security of AES-128 against quantum attacks, requiring the use of longer keys to maintain the same security level.

Post-Quantum Attacks: Beyond encryption-breaking, quantum computers could pose new types of cybersecurity risks by discovering vulnerabilities in software or systems through faster code analysis and reverse engineering techniques. In the future, adversaries may develop sophisticated quantum-based cyberattacks that traditional defenses cannot detect or counter.

Building Quantum-Resistant Solutions

To protect data in the quantum era, cybersecurity experts are developing a range of quantum-resistant cryptographic methods and technologies. These post-quantum solutions aim to safeguard data against both classical and quantum computers, creating a resilient defense for the future.

Post-Quantum Cryptography (PQC): Post-quantum cryptography refers to encryption algorithms that are designed to be secure against both classical and quantum computing attacks. Unlike quantum cryptography, which uses quantum states, post-quantum cryptography is implemented on classical systems, making it easier to integrate into existing infrastructure.

Lattice-Based Cryptography: Lattice-based cryptography is one of the most promising candidates for post-quantum encryption. It relies on mathematical problems, such as the shortest vector problem (SVP) in a high-dimensional lattice, that are believed to be hard for both classical and quantum computers to solve. Algorithms like Learning with Errors (LWE) and Ring-LWE offer strong security guarantees and are being considered for widespread adoption.

Code-Based Cryptography: Code-based cryptosystems, like the McEliece algorithm, use error-correcting codes to ensure secure communication. These methods have been studied for decades and are resistant to quantum attacks due to the difficulty of decoding randomly generated linear codes.

Multivariate Quadratic Equations: Another post-quantum approach involves using multivariate polynomial equations, which are computationally hard to solve for both classical and quantum machines. Algorithms based on solving these systems offer potential solutions for quantum-resistant encryption.

Hash-Based Signatures: Hash-based signature schemes, such as the Merkle signature scheme, rely on the security of cryptographic hash functions, which are believed to be resistant to quantum attacks. These schemes are particularly useful for creating secure digital signatures in a quantum world.

Quantum Key Distribution (QKD): As discussed in the previous chapter, QKD offers a way to securely distribute cryptographic keys by leveraging the principles of quantum mechanics. While not a replacement for encryption, QKD ensures that any eavesdropping on key exchanges will be detected, providing a level of security that cannot be compromised by quantum computers.

Hybrid Systems: As quantum technologies evolve, many cybersecurity experts advocate for hybrid systems that combine classical and quantum methods. For instance, integrating post-quantum cryptography with QKD can offer both long-term security and immediate protection against potential quantum threats. These systems could provide a transition period while full-scale quantum solutions are developed.

Quantum-Safe Infrastructure: Beyond cryptography, cybersecurity in the quantum era requires rethinking network infrastructure, authentication protocols, and data storage systems. Emerging technologies like quantum random number generators (QRNG) are being explored to improve the randomness and unpredictability of cryptographic keys. This added layer of security ensures keys are as strong as possible, resistant to both classical and quantum attacks.

New Paradigms in Cyber Defense

The quantum revolution will also influence how we approach cyber defense strategies, reshaping them in ways we have yet to fully understand. Here are some key trends:

Quantum-Enhanced Cybersecurity: Quantum computers themselves can also be used as tools for cybersecurity. By leveraging quantum algorithms, defenders can enhance pattern recognition, anomaly detection, and threat prediction systems. This might lead to more powerful intrusion detection systems capable of identifying complex cyberattacks in real-time.

Quantum Forensics: Quantum-enhanced data analysis could be used in digital forensics to trace the origins of cyberattacks, decrypt encrypted malicious payloads, and analyze vast amounts of network data to pinpoint vulnerabilities. This quantum advantage may allow security teams to react faster to incidents and prevent future attacks.

AI-Driven Quantum Cybersecurity: The convergence of artificial intelligence (AI) and quantum computing may create highly sophisticated defense systems that can predict and respond to attacks autonomously. Quantum machine learning models could analyze cyber threats and adapt to new attack vectors more effectively than classical systems, reducing response times and minimizing damage.

Zero-Trust Architectures: In a quantum future, the concept of zero-trust architectures may become more critical. In this model, trust is never assumed within a network, and continuous verification is required for access to any system. By incorporating post-quantum cryptographic methods, zero-trust architectures could offer an additional layer of protection in a world where quantum computers might be used by malicious actors.

Preparing for the Quantum Era

Preparing for the quantum era of cybersecurity requires a proactive approach. Governments, industries, and security researchers are already working toward building quantum-resilient systems, but widespread readiness will take time. Key steps for preparing for this new age include:

Risk Assessment: Organizations must evaluate their current encryption systems and assess the potential risks quantum computing poses. Identifying which systems are most vulnerable and planning for their eventual replacement or upgrade is critical.

Adoption of Post-Quantum Algorithms: Cybersecurity infrastructure will need to adopt quantum-resistant cryptography as part of a long-term strategy. Governments and standards bodies, like the National Institute of Standards and Technology (NIST), are already working to standardize post-quantum algorithms, but early adoption will be crucial for high-security industries like finance, defense, and healthcare.

Quantum Awareness Training: A key aspect of preparing for the quantum era is ensuring that cybersecurity professionals are aware of the threats posed by quantum computing and understand how to implement quantum-resistant technologies. Training programs focused on quantum cryptography and post-quantum solutions will help equip organizations to face this evolving threat.

Collaboration Across Industries: Quantum cybersecurity will require collaboration between industries, governments, and research institutions. By sharing knowledge, resources, and best practices, the global community can collectively ensure the security of data and systems in the quantum future.

CHAPTER 8:

Quantum and AI: A New Partnership

Artificial intelligence (AI) and machine learning (ML) have already revolutionized industries by automating complex tasks, analyzing vast data sets, and enabling systems to learn from experience. However, classical computing faces limitations when it comes to processing the immense amounts of data required for advanced AI applications and solving the increasingly complex problems posed by these systems. Quantum computing offers a way to overcome these challenges, promising to exponentially enhance the capabilities of AI and ML.

In this chapter, we will explore how quantum computing will fuel advancements in AI, making systems smarter, faster, and more efficient. We will dive into the symbiotic relationship between quantum computing and AI, the potential breakthroughs in machine learning, and the transformative impact this partnership will have across industries.

The Limits of AI on Classical Computers

Before understanding the synergy between quantum computing and AI, it's important to grasp the challenges AI faces in the classical computing world.

Data Explosion: AI systems require large amounts of data to train algorithms and improve accuracy. Processing and analyzing this data on classical computers is computationally expensive and time-consuming. In some cases, the sheer volume of data is too vast for classical systems to handle efficiently.

Complexity of Algorithms: Machine learning algorithms, especially deep learning models, are becoming increasingly complex. Training these models requires enormous computational power, and even with advances in GPU and cloud computing, training AI systems can take days or weeks.

Optimization Problems: AI is used to solve a wide range of optimization problems, such as route planning, resource allocation, and financial modeling. Classical algorithms struggle with these problems, particularly when the number of variables and constraints grows exponentially, making them nearly impossible to solve in a reasonable timeframe.

Energy Consumption: AI and ML models require significant computational resources, leading to high energy consumption. Quantum computing promises more efficient solutions that could reduce the environmental impact of AI processing.

Quantum Computing: A New Era for AI

Quantum computing's ability to process information in parallel and solve complex problems faster than classical computers makes it a perfect partner for AI. By leveraging the principles of superposition, entanglement, and quantum parallelism, quantum computing can significantly enhance AI systems in several key ways:

Faster Data Processing: Quantum computers can process data exponentially faster than classical computers. This speed can dramatically reduce the time required to train machine learning models, enabling AI systems to learn from larger data sets in a fraction of the time. For instance, training a neural network that takes weeks on a classical computer could potentially be done in hours or even minutes on a quantum machine.

Solving Complex Optimization Problems: Quantum computers excel at solving optimization problems that classical computers find intractable. This capability could be used to optimize neural network architectures, find the most efficient paths in complex systems, or develop better AI algorithms.

Efficient Pattern Recognition: Quantum computers can analyze massive amounts of data and detect patterns more efficiently than classical systems. In fields like healthcare, quantum-enhanced AI could quickly identify correlations in genomic data, improving disease diagnosis and personalized treatment plans.

Enhanced Machine Learning Models: Quantum computing opens up new possibilities for developing advanced machine learning models. Quantum algorithms, such as the Quantum Approximate Optimization Algorithm (QAOA) and Quantum Support Vector Machines, offer novel approaches to classification, clustering, and optimization problems that underpin many AI applications.

Quantum Machine Learning (QML): A Game Changer

Quantum Machine Learning (QML) is the emerging field that blends the principles of quantum computing with machine learning techniques. QML is poised to revolutionize AI by offering entirely new frameworks for learning and prediction. Several areas of AI stand to benefit from quantum-enhanced machine learning.

Quantum Neural Networks (QNNs): Quantum neural networks (QNNs) are quantum-inspired versions of classical neural networks. While traditional neural networks rely on classical bits, QNNs utilize qubits, allowing for massively parallel computation. This could result in faster training times and more efficient learning models.

Improved Learning Speed: With QNNs, training times could be reduced significantly, as quantum computers can process multiple possibilities simultaneously. This could enable real-time learning in AI applications, making adaptive systems more responsive and intelligent.

Complex Problem Solving: Quantum neural networks could excel in solving highly complex problems, such as those found in drug discovery, where the number of potential molecular combinations is astronomically high. QNNs can rapidly explore many configurations, identifying the most promising solutions.

Quantum Support Vector Machines (QSVMs): Support Vector Machines (SVMs) are widely used in classical machine learning for classification tasks. Quantum versions of SVMs—QSVMs—can handle much larger data sets with higher dimensions. By utilizing quantum states to represent complex data, QSVMs could drastically improve the accuracy and speed of classification problems in areas like image recognition, speech processing, and natural language understanding.

Quantum-Enhanced Optimization: Many AI tasks involve optimization, such as minimizing error functions in neural networks or optimizing decision-making strategies in AI systems. Quantum algorithms, such as the Quantum Approximate Optimization Algorithm (QAOA), can find near-optimal solutions to these problems far more efficiently than classical methods. This has profound implications for industries like finance, logistics, and energy management.

Quantum Generative Models: Generative models, such as Generative Adversarial Networks (GANs), are used to create new data (e.g., images, text, or music) based on existing patterns. Quantum generative models take advantage of quantum superposition to generate complex data patterns that classical models might struggle to create. This could lead to more sophisticated content generation in creative industries or more accurate simulations in scientific research.

Real-World Applications of Quantum-Enhanced AI

The partnership between quantum computing and AI has the potential to reshape industries and create breakthroughs in several fields:

Healthcare and Drug Discovery: AI is already being used to analyze medical data, identify disease patterns, and accelerate drug discovery. Quantum-enhanced AI could take these applications to the next level. By quickly processing vast amounts of genomic, clinical, and molecular data, quantum AI could help develop personalized treatment plans, discover new drug compounds, and even simulate the behavior of biological systems at a molecular level.

Financial Services: In the financial sector, AI is used for fraud detection, risk management, and algorithmic trading. Quantum computing could revolutionize these applications by enabling faster, more accurate simulations of financial markets, optimizing portfolios in real-time, and detecting fraud patterns that are invisible to classical systems.

Supply Chain Optimization: Quantum-enhanced AI could be a game changer for logistics and supply chain management. By solving complex optimization problems, such as route planning, resource allocation, and demand forecasting, quantum AI can improve efficiency and reduce costs in global supply chains.

Natural Language Processing (NLP): AI-driven language models, such as OpenAI's GPT series, have made significant advances in understanding and generating human language. Quantum computing could enhance NLP models by allowing them to process and understand more complex language structures, leading to more accurate translations, better voice assistants, and more advanced text analysis for research and communication.

Climate Modeling and Energy Management: AI is increasingly used to predict climate patterns, model weather systems, and optimize energy consumption. Quantum-enhanced AI could simulate highly complex environmental systems more accurately, enabling scientists to predict climate change scenarios and develop sustainable solutions for energy consumption.

Autonomous Systems: From self-driving cars to drones, autonomous systems rely on AI to navigate complex environments and make real-time decisions. Quantum computing could enhance the decision-making capabilities of these systems, allowing them to process sensor data more quickly, optimize routes more efficiently, and improve safety and reliability.

Challenges and the Road Ahead

While the partnership between quantum computing and AI holds immense promise, several challenges must be overcome:

Quantum Hardware Limitations: The development of large-scale, fault-tolerant quantum computers is still in progress. Quantum systems are currently limited in terms of qubit count and coherence time, which constrains their ability to tackle real-world AI problems. However, as quantum hardware improves, the potential of quantum-enhanced AI will continue to grow.

Algorithm Development: Developing quantum algorithms that can effectively solve AI-related tasks is a significant challenge. Quantum algorithms for machine learning are still in the early stages, and it will take time for researchers to refine these methods and demonstrate their superiority over classical approaches.

Integration with Classical AI Systems: While quantum computing offers significant advantages, most AI systems today run on classical computers. Integrating quantum AI solutions into existing infrastructures will require a hybrid approach, where classical and quantum systems work together. This will involve developing frameworks and tools to bridge the gap between the two paradigms.

Talent and Expertise: Quantum computing and AI are both highly specialized fields, and there is a shortage of professionals with expertise in both areas. Building a workforce capable of developing quantum-enhanced AI solutions will require educational initiatives, interdisciplinary collaboration, and significant investment in training.

CHAPTER 9:

The Quantum Internet: Reimagining Connectivity

The internet as we know it has revolutionized communication, commerce, and culture. However, the classical internet is also fraught with vulnerabilities, from data breaches to cyberattacks, and faces limitations in speed, security, and scalability. Enter the quantum internet—an emerging concept that has the potential to fundamentally transform the way we connect, share, and communicate in the digital world.

The quantum internet, powered by the principles of quantum mechanics, promises a new era of unhackable networks, instant information transfer, and unparalleled security in data sharing. In this chapter, we explore what the quantum internet is, how it differs from the classical internet, the technological advancements enabling it, and its implications for the future of communication.

The Classical Internet: Strengths and Weaknesses

Before diving into the quantum internet, it's essential to understand the current state of the classical internet, which is built on classical bits—binary 1s and 0s. The classical internet connects billions of devices, enabling seamless communication and data exchange across the globe. It has empowered industries, democratized access to information, and driven technological innovation at an unprecedented scale.

However, despite its strengths, the classical internet has several key limitations:

Security Vulnerabilities: Current encryption methods, such as RSA and elliptic curve cryptography, rely on mathematical problems that classical computers struggle to solve. However, as discussed in earlier chapters, these methods are vulnerable to quantum attacks once powerful quantum computers are developed.

Latency and Bandwidth Constraints: The classical internet faces challenges in reducing latency (the time it takes for data to travel between devices) and increasing bandwidth (the amount of data that can be transferred at a given time). As the volume of data being transmitted globally grows exponentially, these limitations become more pronounced.

Data Integrity and Trust: Ensuring the authenticity and integrity of data is a major concern. Currently, it's possible to intercept, modify, or spoof data as it travels across networks, leading to a range of security risks, from fraud to identity theft.

Centralization of Control: The classical internet is largely centralized, with data often passing through large servers or data centers controlled by corporations or governments. This centralization creates potential single points of failure, which can be targeted in cyberattacks or disrupted by outages.

What is the Quantum Internet?

The quantum internet is a proposed global network that would allow quantum computers to communicate with one another using the principles of quantum mechanics. Unlike the classical internet, which transmits data in binary bits, the quantum internet transmits data using quantum bits, or qubits. These qubits leverage the quantum properties of superposition and entanglement to create a fundamentally different mode of communication.

Qubits: Unlike classical bits, which are either 0 or 1, qubits can exist in a superposition of both states at the same time. This allows quantum computers to perform many calculations simultaneously, enabling faster data processing.

Quantum Entanglement: One of the key features of the quantum internet is the use of quantum entanglement, a phenomenon where two or more qubits become intertwined in such a way that the state of one qubit instantaneously affects the state of another, even over long distances. This "spooky action at a distance," as Einstein famously called it, enables instantaneous and secure communication.

Quantum Teleportation: Quantum teleportation is a method of transferring quantum states from one qubit to another, regardless of the physical distance between them. This process is essential for building a quantum internet, as it allows for the transfer of information without the need to send physical particles across a network.

Quantum Key Distribution (QKD): One of the earliest and most practical applications of the quantum internet is quantum key distribution (QKD). QKD allows for the secure exchange of cryptographic keys using quantum mechanics. Any attempt to intercept the key would immediately disturb the quantum system, alerting both parties to the eavesdropping and ensuring that the key exchange is unhackable.

Key Technologies Enabling the Quantum Internet

While the quantum internet is still in its infancy, several key technologies are being developed to make it a reality:

Quantum Repeaters: One of the challenges in building a quantum internet is the distance over which quantum information can be transmitted. Qubits are extremely delicate and can easily lose their quantum state through a process known as "decoherence" when transmitted over long distances. To overcome this, researchers are developing quantum repeaters, devices that can extend the range of quantum communication by amplifying and correcting quantum information without disturbing the quantum state.

Quantum Networks: The quantum internet will require a new kind of network infrastructure that can transmit and manage quantum states. Quantum networks use optical fibers and quantum satellites to enable communication between quantum devices. Quantum satellites, like the Chinese satellite Micius, have already demonstrated successful quantum entanglement and QKD over hundreds of kilometers, laying the groundwork for larger-scale quantum networks.

Quantum Memory: Quantum memory is essential for storing and retrieving quantum information over time. Unlike classical memory, which stores bits of data, quantum memory must be able to preserve the delicate quantum state of qubits without decoherence. This technology is crucial for creating reliable quantum networks, as it allows quantum data to be stored and transmitted securely.

Quantum Routers: Quantum routers will be required to direct quantum information across networks, just like classical routers do today. However, quantum routers must be able to handle the complex nature of quantum states without disturbing the data, making this a challenging yet vital technology for the development of the quantum internet.

The Benefits of the Quantum Internet

The quantum internet has the potential to revolutionize many aspects of communication and data sharing, offering several distinct advantages over the classical internet:

Unhackable Security: The most significant advantage of the quantum internet is its unparalleled security. Quantum communication protocols, such as QKD, make it impossible for eavesdroppers to intercept information without being detected. Any attempt to observe or tamper with quantum data would immediately alter its state, alerting users to the presence of a breach. This unhackable security could protect everything from financial transactions to national security communications.

Instantaneous Information Transfer: Through the phenomenon of quantum entanglement, the quantum internet could enable instantaneous information transfer across vast distances. Unlike classical communication, which is limited by the speed of light, quantum entanglement allows for the transfer of information without the need to physically send particles across space. This has profound implications for applications that require real-time data, such as autonomous vehicles, remote surgery, and global financial markets.

Secure Data Sharing: As more devices become interconnected through the Internet of Things (IoT), the need for secure data sharing will become increasingly critical. The quantum internet could provide a secure infrastructure for sharing sensitive data between devices, ensuring that data integrity and privacy are maintained.

Decentralization: Unlike the classical internet, which relies on centralized servers, the quantum internet could be decentralized. Quantum networks could use peer-to-peer communication models, reducing the risk of single points of failure and giving users more control over their data.

Applications of the Quantum Internet

While the quantum internet is still under development, several potential applications are already being explored:

Financial Transactions and Cryptography: The financial industry relies heavily on secure communication and encryption to protect transactions and personal information. The quantum internet could revolutionize banking by providing unhackable communication channels, ensuring that sensitive data is never intercepted or altered. Quantum cryptography could also be used to safeguard blockchain systems, enhancing the security of cryptocurrencies and digital assets.

Government and Military Communications: Secure communication is critical for government and military operations. The quantum internet could provide a foolproof method for transmitting classified information, ensuring that sensitive communications remain secure from espionage and cyberattacks.

Healthcare and Telemedicine: In healthcare, the quantum internet could be used to securely share patient data between hospitals, clinics, and research institutions. It could also enable real-time, secure communication for telemedicine, allowing doctors to remotely perform procedures or monitor patients with complete privacy.

Autonomous Systems: Quantum communication could enhance the reliability and security of autonomous systems, such as self-driving cars, drones, and robotics. These systems rely on real-time data to make decisions, and the quantum internet could provide a secure, instantaneous communication network to ensure their safety and efficiency.

Global Research Collaboration: The quantum internet could accelerate scientific discovery by enabling global collaboration between researchers. Quantum communication could facilitate the sharing of experimental data in real time, even across continents, leading to faster progress in fields like physics, biology, and medicine.

Challenges to Building the Quantum Internet

While the quantum internet holds immense promise, there are several challenges that must be overcome before it becomes a reality:

Technical Complexity: Quantum systems are highly sensitive and prone to decoherence, making it difficult to transmit quantum information over long distances without losing data integrity. Developing robust quantum repeaters, routers, and memory systems will be crucial for ensuring the reliability of quantum networks.

Infrastructure Development: Building a global quantum internet will require entirely new infrastructure, from quantum communication networks to quantum satellites. This infrastructure will need to be integrated with existing classical networks, requiring significant investment and collaboration between governments, industries, and research institutions.

Cost: The cost of developing quantum hardware and networks is currently high, limiting the widespread deployment of quantum internet technology. However, as quantum technologies mature and become more accessible, the cost of building and maintaining quantum networks is expected to decrease.

Interoperability: For the quantum internet to be successful, it must be able to interoperate with classical networks and devices.

Developing protocols that allow quantum and classical systems to communicate seamlessly will be essential for integrating the quantum internet into existing infrastructures.

The Future of the Quantum Internet

The quantum internet is still in its early stages, but its potential to reshape global communication is undeniable. As research progresses and technology advances, we may soon see the first quantum networks being deployed, offering unprecedented security, speed, and efficiency in communication.

The quantum internet represents the next evolutionary leap in connectivity, a system that will not only enhance our current digital infrastructure but will also unlock new possibilities for innovation in fields ranging from finance and healthcare to science and national security. As the quantum revolution unfolds, the quantum internet will become a foundational element of the quantum society, reimagining how we connect and share information in a secure, interconnected world.

CHAPTER 10:

Quantum Medicine: The Future of Health

The healthcare industry is on the verge of a profound transformation, driven by the power of quantum computing. Today, medical advancements rely on classical computing technologies to process data, analyze genetic codes, and model biological systems. However, classical computers are increasingly hitting their limits when it comes to solving the complex, high-dimensional problems that define modern medicine, such as drug discovery, disease diagnosis, and personalized treatment.

Quantum computing offers an unprecedented opportunity to break through these barriers. By harnessing the unique properties of quantum mechanics, quantum computers can process vast amounts of biological data, simulate molecular interactions with unparalleled accuracy, and solve optimization problems that classical computers struggle with. In this chapter, we will explore how quantum computing is poised to revolutionize healthcare—from accelerating drug discovery to enabling personalized medicine and quantum simulations of biological systems.

The Challenges of Healthcare Today

Healthcare faces numerous challenges that hinder the delivery of effective treatments and personalized care:

Slow and Costly Drug Discovery: Developing new drugs can take more than a decade and billions of dollars in investment. Identifying viable drug compounds and testing their effects on the human body through laboratory experiments and clinical trials is an incredibly time-consuming and expensive process.

Complex Biological Systems: Human biology is exceedingly complex, involving interactions between trillions of cells, molecules, and genes. Modeling and simulating these interactions on classical computers is an immense challenge, particularly when it comes to understanding diseases like cancer or neurodegenerative disorders.

One-Size-Fits-All Treatments: While medical advancements have saved millions of lives, many treatments remain "one-size-fits-all" approaches that fail to account for individual differences in genetics, lifestyle, and environmental factors. Personalized medicine—tailoring treatments to an individual's unique biology—requires the ability to process and analyze large-scale genetic data, which classical computing often struggles to handle.

Diagnosis Challenges: Many diseases are difficult to diagnose accurately in their early stages, often leading to delayed treatments or misdiagnoses. Advanced diagnostic tools that can analyze complex medical data, such as imaging or genetic information, are needed to detect diseases with greater precision.

Quantum computing holds the potential to address these challenges by delivering solutions that classical computers cannot. Let's explore how this technology can unlock new possibilities for medical innovation.

Faster and More Efficient Drug Discovery

One of the most promising applications of quantum computing in healthcare is drug discovery. Currently, drug discovery is an expensive and slow process, with pharmaceutical companies needing to sift through millions of potential compounds to find the one that works effectively and safely in humans. This process involves a lot of trial and error, often requiring years of testing in labs, animal models, and clinical trials.

Quantum computing can significantly accelerate this process by simulating the behavior of molecules and proteins at the quantum level. This would allow scientists to predict how different compounds will interact with biological targets (such as enzymes or receptors) without the need for exhaustive laboratory testing. With quantum computers, researchers can explore a vast chemical space in a fraction of the time, drastically reducing the time it takes to identify promising drug candidates.

Molecular Simulation and Quantum Chemistry: Quantum computers are uniquely suited for molecular simulations because they can model the behavior of subatomic particles, such as electrons, more accurately than classical computers. These simulations allow researchers to understand how molecules interact with each other and with biological systems on a fundamental level.
For example, researchers could use quantum computers to simulate the interaction between a potential drug compound and a protein associated with a specific disease. By understanding this interaction at the quantum level, scientists can predict whether the compound will bind effectively to the target protein, potentially inhibiting its harmful effects.

Case Study: Simulating Antibiotic Resistance: One real-world example involves using quantum simulations to address antibiotic resistance. Bacteria are evolving to resist existing antibiotics, creating a major global health crisis. Quantum simulations could help researchers design new antibiotics by predicting how bacterial proteins will mutate and develop resistance to drugs. This would enable scientists to develop antibiotics that can stay ahead of bacterial evolution.

Optimizing Drug Combinations: In many cases, the best treatment for a disease involves a combination of drugs. However, finding the optimal combination is a complex optimization problem. Quantum computers excel at solving such problems by exploring multiple variables simultaneously and identifying the best solution. This could help doctors personalize drug regimens for individual patients, improving treatment outcomes while minimizing side effects.

Personalized Medicine: Tailoring Treatments to Individuals

Personalized medicine aims to move beyond the "one-size-fits-all" approach and tailor treatments to a patient's unique genetic makeup, environment, and lifestyle. This approach relies on large-scale analysis of genomic, proteomic, and clinical data to determine the most effective treatment for each individual.

However, personalized medicine generates an enormous amount of data, particularly when it comes to analyzing a person’s genome. The human genome consists of approximately 3 billion base pairs, and understanding how genetic variations contribute to disease requires analyzing complex, high-dimensional datasets. Quantum computing offers the computational power needed to process this data more efficiently, leading to breakthroughs in personalized medicine.

Genetic Analysis and Quantum Machine Learning: Quantum computing can dramatically speed up the analysis of genetic data by leveraging quantum machine learning algorithms. These algorithms can sift through genomic data, identifying patterns and correlations that classical algorithms may miss. By analyzing how specific genetic mutations affect an individual's response to certain drugs, quantum-enhanced AI systems could help doctors develop customized treatment plans.

For example, cancer treatment often involves chemotherapy, but not all patients respond to the same drugs in the same way. By analyzing the patient's genetic profile and comparing it with data from other patients, quantum algorithms could predict which chemotherapy regimen is most likely to be effective, reducing trial-and-error approaches.

Tailoring Immunotherapies: Immunotherapy is a cutting-edge approach that uses a patient's immune system to fight diseases like cancer. However, not all patients respond equally to immunotherapy, and understanding why requires analyzing vast amounts of immunological and genetic data. Quantum computers could analyze these data sets to personalize immunotherapy treatments, ensuring that each patient receives the most effective immune-targeting strategies.

Quantum Diagnostics: Better Accuracy, Faster Results

Quantum computing also has the potential to revolutionize diagnostics by enhancing the accuracy and speed of disease detection. By processing complex data from medical imaging, genetic testing, and other diagnostic tools, quantum systems can improve early diagnosis, allowing for earlier interventions and better patient outcomes.

Quantum-Enhanced Medical Imaging: Medical imaging technologies like MRI, CT scans, and X-rays generate detailed images of the body's internal structures. These images are often complex and require advanced algorithms to detect anomalies, such as tumors or blockages. Quantum computing could enhance image analysis, allowing doctors to detect early-stage diseases more accurately and quickly.

For example, in cancer diagnosis, early detection is critical for successful treatment. Quantum-enhanced imaging could help radiologists identify tiny tumors or abnormal tissue growth that might be missed by classical imaging technologies, improving early diagnosis rates.

Quantum Biomarkers for Disease Detection: Quantum sensors can detect subtle changes in the biological environment, such as minute variations in electromagnetic fields or molecular structures. These sensors could be used to develop quantum biomarkers—indicators that signal the presence of a disease at a molecular level. By identifying these biomarkers early, quantum diagnostics could allow for more precise and earlier diagnosis of diseases like Alzheimer's, Parkinson's, and cardiovascular conditions.

Real-Time Data Analysis: Quantum computing could enable real-time analysis of diagnostic data, allowing doctors to make faster, more informed decisions. For example, during surgery, quantum-enhanced monitoring systems could analyze a patient's vital signs and detect potential complications before they become life-threatening, improving the safety and success rates of medical procedures.

Quantum Simulations of Biological Systems

Understanding the complexity of biological systems—whether at the level of individual molecules or entire organisms—is one of the biggest challenges in modern science. Classical computers struggle to simulate biological systems due to their immense complexity and the vast number of interactions involved. Quantum computing, however, is uniquely suited to simulating biological systems at a molecular level.

Simulating Protein Folding: Proteins are essential molecules that play a key role in virtually all biological processes. However, proteins must fold into specific three-dimensional structures to function properly, and misfolded proteins are associated with diseases such as Alzheimer's, Parkinson's, and Huntington's. Predicting how a protein will fold is a highly complex problem that classical computers struggle to solve.

Quantum computers could simulate the quantum forces that drive protein folding, allowing researchers to predict the structure of proteins with greater accuracy. This could lead to breakthroughs in understanding diseases caused by protein misfolding and the development of targeted therapies to prevent or reverse these processes.

Modeling Biological Networks: Quantum computers can simulate entire biological networks, such as neural networks in the brain or metabolic pathways in cells. By modeling these networks, researchers can gain a deeper understanding of how diseases like cancer or neurological disorders develop and progress. These simulations could also be used to test how different treatments affect biological systems, leading to more effective therapies.

Quantum Simulations of Drug Interactions: Quantum simulations can model how different drug compounds interact with various biological targets, such as enzymes, proteins, or DNA. This level of precision allows researchers to identify potential side effects, optimize drug dosages, and predict how different patients will respond to a treatment based on their unique biology.

The Road Ahead: Challenges and Opportunities

While the potential of quantum medicine is immense, several challenges must be overcome before quantum computing becomes a standard tool in healthcare:

Quantum Hardware Development: Quantum computers are still in the early stages of development, with current machines being prone to errors and limited in their number of qubits. Developing fault-tolerant, scalable quantum computers will be essential for unlocking the full potential of quantum medicine.

Data Privacy and Ethics: The use of quantum computing in healthcare raises important questions about data privacy and ethics. With quantum computers able to process vast amounts of personal health data, ensuring that this information is securely protected will be critical.

Integration with Classical Systems: Quantum computers will need to work alongside classical systems in healthcare. Developing hybrid quantum-classical algorithms and systems that can integrate seamlessly with existing medical technologies will be necessary for widespread adoption.

Training and Education: Healthcare professionals will need to be trained in quantum technologies and their applications to medicine. As quantum medicine becomes more mainstream, universities and medical schools will need to incorporate quantum computing into their curricula.

CHAPTER 11:

The Future of Climate Science

Climate change stands as one of the most pressing challenges of our time, posing significant risks to ecosystems, economies, and human health. Despite decades of research and policy efforts, progress in understanding and mitigating climate change has been hindered by the complexity of climate systems and the limitations of classical computing. Quantum computing, with its unprecedented processing power and ability to model intricate systems, offers a transformative approach to climate science. In this chapter, we will explore how quantum computing is poised to revolutionize our understanding of climate change, optimize renewable energy systems, and develop innovative solutions to create a sustainable future.

The Challenge of Climate Change

Climate change is driven by various factors, including greenhouse gas emissions, land-use changes, and industrial activities. The complexity of the climate system, involving intricate interactions between the atmosphere, oceans, land, and living organisms, makes it difficult to predict future climate patterns accurately. Key challenges include:

Complex Climate Models: Traditional climate models rely on mathematical equations that describe the physical processes occurring in the atmosphere and oceans. However, these models often simplify complex interactions, leading to uncertainties in projections of future climate scenarios.

Data Overload: Climate science generates massive amounts of data from satellite observations, climate simulations, and ground measurements. Analyzing this data effectively to inform policy decisions and mitigation strategies is an ongoing challenge.

Renewable Energy Integration: Transitioning to renewable energy sources such as wind, solar, and hydroelectric power is critical for reducing greenhouse gas emissions. However, optimizing these systems to meet energy demands while ensuring stability and efficiency poses significant computational challenges.

Quantum Computing: A Game Changer for Climate Science

Quantum computing can address the challenges of climate science through its unique capabilities, including superposition, entanglement, and quantum parallelism. By leveraging these properties, researchers can model climate systems more accurately, analyze vast datasets efficiently, and optimize renewable energy systems effectively.

Enhanced Climate Modeling: Quantum computers can simulate complex climate systems more accurately than classical computers. Traditional models often rely on approximations, leading to uncertainties in long-term climate predictions. Quantum algorithms can capture the intricacies of climate dynamics, including:

Atmospheric Interactions: Quantum computers can model the interactions between various atmospheric components, such as clouds, aerosols, and greenhouse gases. This improved understanding will lead to better predictions of climate feedback mechanisms, which are crucial for accurate climate projections.

Ocean-Atmosphere Dynamics: The exchange of heat and moisture between the ocean and atmosphere plays a critical role in regulating climate. Quantum computing can simulate these interactions at a higher resolution, allowing scientists to explore the impact of ocean currents on weather patterns and climate variability.

Land Surface Processes: Land use changes, deforestation, and urbanization significantly affect climate. Quantum models can incorporate these factors more effectively, enabling researchers to assess their impacts on local and global climates.

Optimizing Renewable Energy Systems: Transitioning to a sustainable energy future relies on optimizing renewable energy generation, storage, and distribution. Quantum computing can enhance energy systems in several ways:

Grid Management: Quantum algorithms can analyze vast amounts of data from smart grids to optimize energy distribution and load balancing. This will improve the reliability and efficiency of renewable energy sources, enabling a smoother transition away from fossil fuels.

Energy Storage Solutions: Efficient energy storage is vital for managing the intermittent nature of renewable sources like solar and wind. Quantum simulations can help researchers develop advanced battery technologies and other storage solutions by modeling molecular interactions and optimizing material properties.

Site Selection for Renewable Projects: Quantum computing can analyze geographical, environmental, and logistical data to identify optimal locations for wind farms, solar arrays, and other renewable energy projects. By considering multiple variables simultaneously, quantum algorithms can ensure efficient resource allocation.

Data Analysis and Climate Forecasting: Climate scientists rely on vast datasets collected from satellites, sensors, and climate models. Analyzing this data efficiently is critical for informing policy decisions and developing adaptive strategies. Quantum computing can revolutionize data analysis through:

Quantum Machine Learning: Quantum machine learning algorithms can process large datasets more efficiently than classical algorithms, enabling faster insights into climate trends and patterns. These algorithms can detect correlations and anomalies that classical methods may overlook, leading to more accurate climate predictions.

Improved Forecasting Models: By integrating quantum algorithms into climate forecasting, scientists can develop models that provide real-time updates on climate conditions. This capability will enhance disaster preparedness and inform policy responses to extreme weather events.

Case Studies: Quantum Computing in Action

Quantum Simulations for Carbon Capture: Carbon capture and storage (CCS) technologies are essential for mitigating greenhouse gas emissions from industrial processes. Quantum computing can model the molecular interactions involved in capturing carbon dioxide, helping researchers identify more efficient capture materials and processes. For example, a quantum simulation might analyze how different materials interact with CO2 at the atomic level, leading to breakthroughs in the design of more effective capture technologies.

Climate Impact Assessments: Researchers can use quantum algorithms to conduct detailed assessments of how specific policies, such as carbon pricing or renewable energy incentives, will impact climate outcomes. By modeling the interactions between economic, environmental, and social factors, quantum computing can provide policymakers with actionable insights into the efficacy of different strategies.

Modeling Extreme Weather Events: The frequency and intensity of extreme weather events are increasing due to climate change. Quantum computing can enhance the modeling of these events, allowing scientists to understand the underlying processes that contribute to hurricanes, floods, and heatwaves. Improved modeling will enable better preparedness and response strategies, ultimately saving lives and reducing economic impacts.

Ethical Considerations and Challenges

While quantum computing holds great promise for climate science, it also presents challenges and ethical considerations that must be addressed:

Equity in Access to Technology: As quantum computing technology advances, ensuring equitable access to its benefits will be essential. Developing countries may lack the resources to harness quantum computing for climate science, potentially widening the gap in climate resilience and adaptation efforts.

Data Privacy Concerns: The collection and analysis of climate data, particularly when it involves sensitive information about communities or industries, raise privacy concerns. Ensuring data is handled ethically and transparently will be critical as researchers utilize quantum computing for climate-related studies.

Investment in Research and Development: To realize the full potential of quantum computing in climate science, substantial investments in research and development are required. Policymakers, industry leaders, and academic institutions must collaborate to foster innovation and promote the adoption of quantum technologies for climate solutions.

CHAPTER 12:

Transforming Material Science and Chemistry

The quest for new materials has always been at the forefront of scientific discovery, from the invention of plastics to the development of superconductors. As industries and technologies evolve, the demand for materials with specific, often unprecedented properties becomes more pronounced. Traditional methods of material discovery and design are limited by the complexities of atomic and molecular interactions, but quantum computing promises to revolutionize this field. By harnessing quantum simulations, researchers can explore the properties of materials and chemical compounds at an unprecedented scale and accuracy. In this chapter, we will examine how quantum computing is set to transform material science and chemistry, paving the way for breakthroughs that could reshape industries ranging from electronics to pharmaceuticals.

The Challenges of Traditional Material Science

Material science and chemistry face several significant challenges that hinder innovation:

Complex Interactions at the Atomic Level: The behavior of materials arises from interactions at the atomic and molecular levels. Traditional computational methods, based on classical physics, struggle to accurately model these interactions, particularly in complex systems. This limitation can lead to slow progress in discovering new materials with desired properties.

Trial-and-Error Approaches: The traditional approach to material discovery often involves a laborious trial-and-error process. Scientists synthesize various compounds and conduct experiments to test their properties. This method is time-consuming, costly, and often results in missed opportunities to discover materials that could have significant applications.

Data Overload: Advances in experimental techniques generate vast amounts of data related to material properties, compositions, and behaviors. Analyzing this data effectively and extracting meaningful insights requires advanced computational tools that can handle complex datasets.

Quantum Computing: A Paradigm Shift for Material Science and Chemistry

Quantum computing offers a fundamentally different approach to simulating and understanding materials, leveraging the principles of quantum mechanics to tackle challenges that classical computers cannot. Key advantages of quantum computing in material science and chemistry include:

Accurate Molecular Simulations: Quantum computers can simulate molecular interactions with remarkable accuracy, capturing the nuances of electron behavior and quantum states. This ability enables researchers to predict material properties without the need for extensive laboratory testing.

Quantum Chemistry: Quantum chemistry, a field dedicated to understanding molecular behavior using quantum mechanics, can leverage quantum computers to calculate molecular energies, reaction pathways, and potential energy surfaces. This precision allows scientists to explore previously uncharted chemical space.

Molecular Dynamics Simulations: Quantum computing can enhance molecular dynamics simulations, allowing researchers to model the time evolution of complex molecular systems. By accurately predicting how molecules behave over time, scientists can gain insights into reaction mechanisms and material stability.

Discovery of New Materials: Quantum simulations will facilitate the discovery of novel materials with tailored properties, including:

Superconductors: The search for room-temperature superconductors has long been a holy grail in material science. Quantum computing can model the interactions of electrons in complex materials, guiding researchers toward materials that exhibit superconductivity at higher temperatures.

2D Materials: Materials like graphene and transition metal dichalcogenides have unique properties that arise from their two-dimensional structure. Quantum simulations can help identify and design new 2D materials with specific electrical, optical, or mechanical properties, opening up avenues for applications in electronics, photonics, and energy storage.

Polymers and Nanomaterials: Quantum computing can assist in the design of polymers and nanomaterials with specific characteristics, such as enhanced strength, conductivity, or reactivity. By understanding how molecular structure influences properties, researchers can develop materials for applications ranging from drug delivery to advanced manufacturing.

Optimization of Chemical Reactions: Quantum computing can optimize chemical reactions by identifying the most favorable reaction pathways and conditions. This capability can lead to more efficient synthesis methods and reduced environmental impact.

Catalyst Design: Catalysts play a crucial role in chemical reactions, enhancing reaction rates and selectivity. Quantum simulations can guide the design of new catalysts by predicting their effectiveness in promoting specific reactions. This will enable the development of greener and more sustainable chemical processes.

Reaction Networks: Quantum computing can analyze complex reaction networks, allowing researchers to explore how multiple reactions interact with one another. This holistic approach to reaction optimization will accelerate the development of new synthetic routes.

Case Studies: Quantum Computing in Material Discovery

High-Temperature Superconductors: Researchers have long sought materials that can conduct electricity without resistance at room temperature. Quantum simulations can model the interactions between electrons and phonons in candidate materials, helping scientists identify potential high-temperature superconductors. By understanding the underlying mechanisms, researchers can design new materials that exhibit superconductivity under practical conditions.

Hydrogen Storage Materials: Hydrogen fuel cells are a promising technology for clean energy, but effective hydrogen storage remains a challenge. Quantum simulations can help identify materials that can absorb and release hydrogen efficiently, paving the way for advancements in hydrogen storage technologies. Researchers can explore various metal hydrides and metal-organic frameworks (MOFs) to find optimal candidates for hydrogen storage applications.

Pharmaceutical Development: The pharmaceutical industry can significantly benefit from quantum computing. By modeling interactions between drug molecules and biological targets at the quantum level, researchers can design new compounds with improved efficacy and reduced side effects. This approach could accelerate drug discovery timelines and lead to more effective therapies.

Ethical Considerations and Challenges

As quantum computing continues to advance material science and chemistry, several ethical considerations and challenges must be addressed:

Intellectual Property Issues: The discovery of new materials and compounds raises questions about intellectual property rights. As quantum computing accelerates material discovery, it will be crucial to establish frameworks for protecting intellectual property while fostering innovation.

Environmental Impact: While quantum computing has the potential to develop more sustainable materials and processes, it is essential to consider the environmental impact of new materials. Researchers must ensure that newly developed materials are environmentally friendly and do not pose risks to human health or ecosystems.

Access to Quantum Technologies: Ensuring equitable access to quantum computing resources is vital to prevent disparities in material science advancements. Collaboration between academia, industry, and government will be essential to democratize access to quantum technologies and ensure that their benefits are widely shared.

CHAPTER 13:

Quantum Finance: Unleashing New Markets

The financial landscape is undergoing a seismic shift, driven by advancements in technology, data analytics, and now, quantum computing. As the financial sector grapples with increasing complexity, volatility, and competition, traditional methods of analysis and risk management are becoming less effective. Quantum finance emerges as a revolutionary force capable of optimizing investment strategies, predicting market behavior with unprecedented accuracy, and transforming risk management models. In this chapter, we will explore how quantum computing is poised to reshape the financial industry, creating new markets and enhancing the decision-making processes of financial institutions.

The Challenges of Traditional Finance

The traditional financial sector faces numerous challenges that hinder optimal performance and risk management:

Complex Portfolio Optimization: Portfolio management involves selecting the right mix of assets to maximize returns while minimizing risk. However, as the number of assets and variables increases, traditional optimization algorithms struggle to identify optimal combinations efficiently. The computational complexity of these problems often leads to suboptimal investment strategies.

Market Volatility and Predictability: Financial markets are notoriously volatile, influenced by a myriad of factors ranging from economic indicators to geopolitical events. Traditional predictive models, often based on historical data, struggle to account for sudden market shifts and emerging trends. As a result, investors may face unexpected losses or missed opportunities.

Risk Assessment and Management: Accurately assessing and managing risk is essential for financial institutions. However, conventional risk models often rely on simplified assumptions that can overlook critical factors. This limitation can lead to inadequate risk assessments and insufficient safeguards against market downturns.

Quantum Computing: A Game Changer for Financial Systems

Quantum computing offers a unique approach to addressing the challenges faced by the financial sector. By leveraging the principles of quantum mechanics, financial institutions can harness the power of quantum algorithms to enhance their analytical capabilities and decision-making processes.

Optimizing Investment Portfolios: Quantum computing can significantly improve portfolio optimization by enabling the evaluation of an exponentially larger set of potential asset combinations.

Quantum Algorithms for Portfolio Optimization: Algorithms such as the Quantum Approximate Optimization Algorithm (QAOA) can efficiently explore complex investment scenarios, allowing investors to identify optimal asset allocations that maximize returns while minimizing risk. This capability is particularly valuable in managing large portfolios with diverse asset classes.

Real-Time Adjustments: Quantum computing can process vast amounts of real-time market data to facilitate dynamic portfolio adjustments. This responsiveness allows financial institutions to capitalize on emerging market trends and mitigate potential losses.

Predicting Market Behavior: The ability to predict market movements is crucial for traders and investors. Quantum computing can enhance predictive models by analyzing complex data patterns and identifying correlations that may be difficult to detect using classical methods.

Quantum Machine Learning: Quantum machine learning algorithms can analyze historical and real-time market data to uncover hidden patterns and trends. By leveraging quantum algorithms, financial analysts can improve the accuracy of market predictions, enabling more informed trading decisions.

Sentiment Analysis: Analyzing social media sentiment and news articles has become an essential tool for predicting market behavior. Quantum computing can enhance sentiment analysis by processing vast amounts of unstructured data quickly, providing investors with insights into market sentiment and potential price movements.

Transforming Risk Management Models: Effective risk management is critical for financial institutions to protect against potential losses. Quantum computing offers the potential to revolutionize risk assessment and management.

Advanced Risk Modeling: Quantum algorithms can facilitate the development of more sophisticated risk models that account for a broader range of variables and complex interdependencies. This capability enables financial institutions to assess risk more accurately and implement robust mitigation strategies.

Scenario Analysis and Stress Testing: Quantum computing allows for comprehensive scenario analysis and stress testing by simulating various market conditions and their potential impacts on portfolios. This capability helps financial institutions prepare for extreme events and enhance their resilience.

Case Studies: Quantum Computing in Action

Hedge Fund Innovations: Some hedge funds are already experimenting with quantum algorithms to enhance their trading strategies. By leveraging quantum computing to analyze large datasets and identify profitable trading opportunities, these funds aim to outperform traditional strategies and capture alpha in competitive markets.

Risk Management in Banking: Major banks are exploring the use of quantum computing for risk management. By developing quantum algorithms that can analyze complex portfolios and assess risk exposure more accurately, banks can better manage their capital reserves and optimize their regulatory compliance strategies.

Financial Modeling for Startups: Startups focused on fintech are leveraging quantum computing to create innovative financial models. By using quantum simulations to predict market behavior and assess investment opportunities, these startups can develop data-driven solutions that provide a competitive edge in the crowded fintech landscape.

Ethical Considerations and Challenges

While quantum finance holds tremendous promise, it also raises ethical considerations and challenges that must be addressed:

Data Privacy and Security: Financial institutions handle sensitive personal and financial data. As quantum computing advances, ensuring the privacy and security of this data becomes paramount. Developing secure quantum cryptographic methods will be essential to protect against potential breaches.

Access and Equity: The disparity in access to quantum computing resources may exacerbate inequalities in the financial sector. Ensuring that smaller firms and emerging markets have access to quantum technologies will be critical to preventing a concentration of power among a few dominant players.

Regulatory Frameworks: As quantum finance evolves, regulatory frameworks must adapt to address the unique challenges and risks associated with quantum technologies. Policymakers need to collaborate with industry leaders to develop guidelines that ensure responsible and transparent use of quantum computing in finance.

CHAPTER 14:

Supercharging Supply Chains

In an increasingly interconnected world, the complexity of global supply chains has grown exponentially. Companies face a myriad of challenges, from managing inventory to optimizing transportation routes and dealing with disruptions caused by unforeseen events. Traditional supply chain management practices often struggle to adapt to these dynamic conditions, resulting in inefficiencies, higher costs, and delayed deliveries. Enter quantum computing: a transformative technology poised to revolutionize how businesses manage and optimize their supply chains. This chapter explores how quantum computing can supercharge supply chains, reduce inefficiencies, and enhance responsiveness to real-time global challenges in logistics and transportation.

The Challenges of Traditional Supply Chain Management

The traditional approach to supply chain management is fraught with challenges that can hinder efficiency and responsiveness:

Complexity of Network Interactions: Supply chains involve numerous interconnected entities, including suppliers, manufacturers, distributors, and retailers. The interactions between these entities are complex and dynamic, making it challenging to optimize the entire network effectively.

Inefficient Inventory Management: Balancing inventory levels to meet demand while minimizing holding costs is a perpetual challenge for businesses. Traditional forecasting methods often rely on historical data, which can be inaccurate, leading to stockouts or excess inventory.

Disruptions and Risk Management: Natural disasters, geopolitical tensions, and pandemics can disrupt supply chains, causing delays and financial losses. Traditional risk management approaches may not be sufficient to mitigate the impacts of these disruptions, leaving companies vulnerable to supply chain failures.

Transportation Optimization: Efficient transportation is critical for timely deliveries, yet optimizing transportation routes can be a complex problem, especially when considering variables such as traffic conditions, fuel costs, and delivery windows.

Quantum Computing: A Paradigm Shift for Supply Chain Optimization

Quantum computing offers innovative solutions to address the challenges faced by supply chains, providing businesses with powerful tools to enhance efficiency and responsiveness.

Complex Network Optimization: Quantum algorithms can tackle the complexity of supply chain networks by simultaneously considering multiple variables and constraints. This capability allows for more effective optimization of the entire supply chain, leading to reduced costs and improved performance.

Quantum Approximate Optimization Algorithm (QAOA): This algorithm can be used to optimize various supply chain problems, including supplier selection, production scheduling, and distribution logistics. By identifying the best combination of decisions, companies can streamline operations and enhance their competitive edge.

Graph Theory Applications: Quantum computing can apply graph theory to model and analyze supply chain networks. This approach allows businesses to visualize and optimize the flow of goods and information across complex networks, leading to more informed decision-making.

Real-Time Inventory Management: Quantum computing enables businesses to respond swiftly to changing market demands by optimizing inventory levels in real time.

Demand Forecasting: Quantum algorithms can analyze vast amounts of data to provide more accurate demand forecasts, allowing companies to align production and inventory levels with real-time market conditions. By reducing forecasting errors, businesses can minimize stockouts and excess inventory.

Dynamic Inventory Allocation: Quantum computing can facilitate dynamic inventory allocation across multiple locations based on real-time demand and supply conditions. This adaptability ensures that products are available where and when they are needed, reducing lead times and improving customer satisfaction.

Enhanced Risk Management: Quantum computing can revolutionize risk assessment and management in supply chains by providing deeper insights into potential vulnerabilities.

Scenario Analysis: Quantum simulations can model various disruption scenarios, enabling businesses to understand how different factors can impact their supply chains. By assessing the resilience of their supply chains under different conditions, companies can develop proactive strategies to mitigate risks.

Supply Chain Resilience: Quantum computing can identify critical nodes in supply chains and suggest strategies for redundancy and diversification. By building more resilient supply chains, companies can minimize the impacts of disruptions and ensure continuity of operations.

Transportation and Logistics Optimization: Efficient transportation is crucial for timely deliveries, and quantum computing can optimize logistics operations by analyzing complex variables.

Route Optimization: Quantum algorithms can optimize transportation routes by considering multiple factors, including traffic patterns, fuel costs, and delivery schedules. This capability enables companies to reduce transportation costs and improve delivery times.

Load Optimization: Quantum computing can optimize load planning by analyzing cargo space and weight distribution across transportation vehicles. By maximizing cargo capacity, businesses can reduce transportation costs and enhance operational efficiency.

Case Studies: Quantum Computing in Supply Chain Management

Manufacturing and Distribution: Leading manufacturing firms are exploring the use of quantum computing to optimize their supply chains. By implementing quantum algorithms to streamline production schedules, manage inventory, and enhance distribution networks, these firms can improve their operational efficiency and reduce costs.

E-commerce and Last-Mile Delivery: E-commerce giants are leveraging quantum computing to enhance last-mile delivery logistics. By optimizing delivery routes and managing inventory across multiple fulfillment centers, these companies can ensure fast and efficient order fulfillment, meeting the growing expectations of consumers.

Agriculture Supply Chains: Agricultural supply chains face unique challenges related to seasonality and perishability. Quantum computing can help optimize the allocation of resources, manage inventory levels, and streamline distribution processes, ultimately improving the efficiency and sustainability of food supply chains.

Ethical Considerations and Challenges

As quantum computing continues to evolve within supply chain management, several ethical considerations and challenges must be addressed:

Data Privacy and Security: Supply chains handle sensitive information related to suppliers, customers, and logistics. Ensuring data privacy and security will be paramount as quantum computing advances, necessitating robust security measures to protect against breaches.

Job Displacement: The adoption of quantum computing may lead to automation and changes in job roles within supply chain management. Companies must consider the potential impact on employment and invest in reskilling and upskilling their workforce.

Equitable Access to Quantum Technology: As quantum computing becomes increasingly influential, ensuring equitable access to quantum technologies for smaller businesses and developing countries will be crucial. Collaboration and partnerships will play a vital role in democratizing access to these powerful tools.

CHAPTER 15:

Quantum in Agriculture: Feeding a Growing World

As the global population continues to swell—projected to reach nearly 10 billion by 2050—the challenge of feeding this expanding populace intensifies. Agriculture faces mounting pressures, including climate change, diminishing arable land, water scarcity, and the need for sustainable practices. Traditional agricultural methods often fall short in addressing these complex issues. However, quantum computing offers innovative solutions to revolutionize agriculture, optimizing crop yields and enhancing food supply chain management. In this chapter, we will explore how quantum computing can address global agricultural challenges and ensure food security for future generations.

The Challenges Facing Global Agriculture

Agriculture today is fraught with challenges that hinder productivity and sustainability:

Decreasing Crop Yields: Traditional farming practices are increasingly inadequate in producing sufficient yields to meet growing demand. Factors such as soil degradation, pest resistance, and unpredictable weather patterns threaten food production.

Resource Scarcity: Water scarcity and limited access to arable land present significant challenges. The agriculture sector is a major consumer of freshwater resources, and as water availability diminishes, the competition for this vital resource intensifies.

Climate Change: Climate change poses a serious threat to agriculture, leading to erratic weather patterns, increased frequency of extreme weather events, and shifting growing seasons. These changes can devastate crop yields and disrupt supply chains.

Supply Chain Inefficiencies: The global food supply chain is complex, involving multiple stakeholders from producers to consumers. Inefficiencies in logistics, storage, and distribution can lead to significant food waste and increased costs.

Quantum Computing: A Catalyst for Agricultural Innovation

Quantum computing has the potential to transform agriculture by addressing these challenges through advanced data analysis, optimization techniques, and simulation capabilities.

Optimizing Crop Yields: Quantum computing can enhance agricultural productivity by enabling precise analysis of various factors that influence crop growth.

Quantum Simulation of Plant Growth: Quantum simulations can model complex biological processes at the molecular level, helping researchers understand how different environmental conditions affect plant growth. This capability enables the development of optimized agricultural practices tailored to specific crops and climates.

Precision Agriculture: By analyzing data from satellite imagery, sensor networks, and climate models, quantum algorithms can provide farmers with insights into optimal planting times, fertilizer application, and irrigation schedules. This precision helps maximize yields while minimizing resource consumption.

Improving Soil Management: Healthy soil is crucial for successful agriculture. Quantum computing can enhance soil management practices by providing detailed analyses of soil health and fertility.

Soil Composition Analysis: Quantum algorithms can analyze soil samples more effectively than traditional methods, identifying nutrient levels, pH, and microbial activity. This information allows farmers to tailor their soil management practices, improving crop health and productivity.

Predictive Analytics for Soil Health: Quantum computing can help predict soil health changes over time based on various environmental factors. By forecasting potential soil degradation, farmers can implement proactive measures to maintain soil quality.

Water Resource Optimization: Efficient water management is vital for sustainable agriculture, particularly in regions facing water scarcity. Quantum computing can optimize irrigation practices to ensure efficient water use.

Smart Irrigation Systems: Quantum algorithms can analyze weather patterns, soil moisture levels, and crop water requirements to develop smart irrigation systems that deliver precise amounts of water when and where it's needed. This technology minimizes water waste while ensuring crops receive adequate hydration.

Water Resource Allocation: Quantum computing can enhance the allocation of water resources across agricultural regions by analyzing competing demands and optimizing distribution. This capability ensures that water is used efficiently and equitably.

Enhancing Supply Chain Management: Improving the efficiency of food supply chains is crucial for reducing waste and ensuring food security. Quantum computing can streamline logistics and distribution processes.

Demand Forecasting: Quantum algorithms can analyze vast datasets to provide more accurate demand forecasts for agricultural products. By predicting market demand, farmers and distributors can optimize production and reduce food waste.

Logistics Optimization: Quantum computing can optimize transportation routes and storage solutions in the food supply chain. By minimizing transportation costs and delivery times, this technology can enhance the efficiency of food distribution, ensuring that fresh produce reaches consumers promptly.

Case Studies: Quantum Computing in Agriculture

Crop Simulation and Yield Prediction: Some agricultural firms are utilizing quantum simulations to model the effects of various environmental factors on crop growth. By accurately predicting yields under different scenarios, these firms can make informed decisions about planting strategies and resource allocation.

Precision Agriculture Innovations: Startups focused on precision agriculture are harnessing quantum computing to analyze data from drones and sensors. By providing real-time insights into soil health, moisture levels, and pest threats, these companies help farmers optimize their operations and improve crop productivity.

Sustainable Water Management: Agricultural cooperatives are exploring quantum computing to enhance water resource management in drought-prone regions. By analyzing weather patterns and soil moisture data, these cooperatives aim to implement efficient irrigation practices that conserve water while maximizing crop yields.

Ethical Considerations and Challenges

While the potential of quantum computing in agriculture is promising, several ethical considerations and challenges must be addressed:

Data Privacy and Security: The use of advanced technologies raises concerns about data privacy, particularly regarding farmers' personal and operational information. Ensuring secure data practices will be critical as quantum computing becomes more prevalent in agriculture.

Equitable Access to Technology: The benefits of quantum computing should be accessible to all farmers, regardless of their size or location. Initiatives must be established to ensure that smallholder farmers and developing regions can access and benefit from quantum technologies.

Environmental Impact: While quantum computing offers the potential to enhance sustainability in agriculture, its implementation must be carefully managed to avoid unintended environmental consequences. Ongoing assessments are necessary to evaluate the long-term impact of quantum applications on ecosystems.

CHAPTER 16:

Quantum Biology: Understanding Life Itself

The mysteries of life, from the intricate mechanisms of cellular processes to the complexities of ecosystems, have long fascinated scientists. Yet, many fundamental biological questions remain unresolved, often due to the limitations of classical computational methods. Quantum biology, an emerging field that intersects quantum mechanics and biological phenomena, offers promising avenues to explore these complexities. In this chapter, we will delve into how quantum computing can revolutionize our understanding of biology, unravel the secrets of life, and tackle some of the most perplexing challenges in the biological sciences.

The Complexity of Biological Systems

Biological systems are incredibly intricate, comprising countless interactions at various scales, from molecules to organisms. Some of the key challenges in understanding these systems include:

Photosynthesis Mechanisms: The process of photosynthesis allows plants to convert sunlight into energy, but the detailed mechanisms behind this process are still not fully understood. The efficiency with which plants capture and convert light is astonishing, often exceeding that of artificial solar cells.

Protein Folding: Proteins are essential to nearly every biological function, and their shape—determined by how they fold—is critical to their function. Misfolded proteins can lead to various diseases, including Alzheimer's and Parkinson's. Accurately predicting protein folding remains one of biology's biggest challenges.

Cellular Behavior: The behavior of cells, including how they respond to environmental stimuli, communicate, and differentiate, involves complex biochemical interactions that are challenging to model accurately. Understanding these behaviors is crucial for advances in medicine and biotechnology.

Quantum Effects in Biological Processes: Some biological processes, such as avian navigation, exhibit behaviors that suggest quantum effects may play a role. Understanding how quantum mechanics influences biology could lead to breakthroughs in multiple fields.

Quantum Computing: A New Paradigm for Biological Research

Quantum computing's unique capabilities make it well-suited for addressing the complexities of biological systems. Here's how quantum computing can facilitate breakthroughs in various areas of biology:

Modeling Photosynthesis: Quantum computing can simulate the intricate processes involved in photosynthesis more effectively than classical computers, potentially revealing the underlying principles that allow plants to harness sunlight so efficiently.

Quantum Simulations of Light Harvesting: By modeling how light-harvesting complexes in plants absorb and transfer energy at the quantum level, researchers can gain insights into optimizing artificial photosynthetic systems. This knowledge could lead to the development of more efficient solar energy technologies.
Advancing Protein Folding Predictions: Quantum computing can tackle the computationally intensive problem of protein folding, potentially enabling researchers to predict how proteins will fold based on their amino acid sequences.

Quantum Algorithms for Protein Folding: Algorithms such as the Quantum Approximate Optimization Algorithm (QAOA) can analyze the energy landscape of protein folding, helping scientists identify the most stable conformations. This capability could accelerate drug discovery by enabling the design of therapeutics that target specific protein shapes.

Understanding Misfolding: Quantum simulations may also provide insights into the conditions that lead to protein misfolding, shedding light on the mechanisms behind various diseases and guiding the development of preventive and therapeutic strategies.

Decoding Cellular Behavior: Quantum computing can enhance our understanding of cellular processes by modeling the complex interactions within cells and their responses to various stimuli.

Quantum Simulations of Cellular Networks: By simulating signaling pathways and metabolic networks at the quantum level, researchers can identify key regulatory mechanisms that dictate cellular behavior. This knowledge can inform the design of targeted therapies for diseases such as cancer, where cellular communication pathways are often disrupted.

Studying Cell-Environment Interactions: Quantum computing can model how cells interact with their environment, including how they respond to changes in nutrient availability, stress, and other factors. This understanding can lead to improved agricultural practices and enhanced crop resilience.

Exploring Quantum Effects in Biology: Some biological phenomena, such as bird navigation and the efficiency of certain enzymes, suggest the presence of quantum effects. Quantum computing can help explore these ideas further.

Avian Navigation: Studies suggest that some birds use quantum entanglement in their navigation systems. By modeling these processes using quantum computing, researchers can gain insights into the mechanisms of avian navigation, potentially leading to advancements in navigation technologies.

Enzyme Efficiency: Quantum computing can also be employed to study how enzymes catalyze reactions with remarkable efficiency. Understanding the quantum properties of these enzymes could inspire the design of more efficient catalysts in industrial processes.

Case Studies: Quantum Computing in Biological Research

Quantum Simulations of Photosynthesis: Researchers have begun to use quantum computers to model the light-harvesting processes in photosynthetic organisms. Early experiments have revealed insights into how energy transfer occurs within chlorophyll molecules, paving the way for more efficient artificial photosynthetic systems.

Protein Folding with Quantum Algorithms: Collaborative projects are exploring the application of quantum algorithms to predict protein folding patterns. Initial results indicate that quantum computing can outperform classical approaches in specific scenarios, providing a promising direction for drug discovery.

Cell Signaling Pathway Analysis: Studies are underway to model complex cell signaling pathways using quantum computing. These efforts aim to identify critical nodes within networks that govern cellular behavior, potentially leading to targeted therapies for diseases linked to signaling dysregulation.

Ethical Considerations and Challenges

As quantum biology advances, several ethical considerations and challenges must be addressed:

Data Privacy and Security: The use of genetic and biological data in quantum research raises concerns about data privacy and security. Researchers must ensure that sensitive information is protected and handled ethically.

Equitable Access to Quantum Technologies: Ensuring equitable access to quantum computing resources for researchers across various disciplines is essential. Collaboration and knowledge sharing will be crucial for democratizing access to these powerful tools.

Environmental Impacts: The application of quantum technologies in biology should be approached with caution to avoid unintended environmental consequences. Rigorous assessments of the ecological impacts of new technologies will be necessary.

CHAPTER 17:

Quantum Simulations: A Window into the Universe

As we stand on the precipice of a new era in scientific exploration, quantum simulations are emerging as a revolutionary tool that allows researchers to delve into the fundamental workings of the universe. By leveraging the unique properties of quantum computing, scientists can simulate complex systems at the quantum level, offering insights into various phenomena in physics, chemistry, and cosmology. This chapter will explore the principles of quantum simulations, their applications across different scientific disciplines, and their potential to reshape our understanding of the universe.

Understanding Quantum Simulations

At its core, quantum simulation involves using quantum computers to model and analyze complex quantum systems. Unlike classical computers, which process information using bits (0s and 1s), quantum computers utilize qubits that can exist in multiple states simultaneously due to superposition. This capability enables quantum computers to perform calculations that are infeasible for classical systems, particularly when dealing with complex quantum interactions.

Superposition and Entanglement:

Superposition allows qubits to represent multiple states at once, enabling simultaneous computations that can explore a vast solution space.
Entanglement links qubits in ways that classical bits cannot, allowing the behavior of one qubit to influence another, regardless of the distance separating them. This property is crucial for simulating correlated quantum systems.
Quantum Algorithms: Various quantum algorithms, such as the Quantum Approximate Optimization Algorithm (QAOA) and the Variational Quantum Eigensolver (VQE), are specifically designed for quantum simulations. These algorithms can efficiently compute properties of quantum systems and find solutions to complex problems.

Applications in Physics

Quantum simulations have profound implications for various fields of physics, allowing researchers to investigate phenomena that classical methods struggle to analyze.

Understanding Quantum Many-Body Systems:

Quantum many-body systems, such as fermions and bosons, exhibit complex behaviors that arise from the interactions between numerous particles. Traditional simulations often fail to capture these dynamics accurately due to the exponential scaling of complexity.

Quantum computers can efficiently model these systems, providing insights into phenomena like superconductivity and magnetism. For instance, researchers have simulated the behavior of electrons in high-temperature superconductors, leading to a better understanding of their underlying mechanisms.

Exploring Quantum Phase Transitions:

Quantum phase transitions occur at absolute zero temperature when a system undergoes a fundamental change in its ground state due to quantum fluctuations.

Quantum simulations can model these transitions, helping physicists explore exotic phases of matter, such as topological insulators and quantum spin liquids.

Investigating Fundamental Particles:

Quantum simulations can aid in understanding the behavior of fundamental particles, such as quarks and gluons, which are the building blocks of protons and neutrons. By modeling interactions in quantum chromodynamics (QCD), researchers can gain insights into the properties of nuclear matter.

Applications in Chemistry

Quantum simulations are transforming the field of chemistry by enabling precise modeling of molecular systems and reactions.

Molecular Modeling:

Quantum computers can accurately simulate the behavior of molecules at the quantum level, providing insights into chemical bonding, reaction pathways, and molecular properties.
For example, researchers have used quantum simulations to study complex molecules, such as drug candidates and catalysts, enabling more efficient drug discovery and material design.

Predicting Reaction Mechanisms:

Understanding the mechanisms of chemical reactions is essential for advancing fields like catalysis and synthetic chemistry. Quantum simulations can help elucidate the intricate pathways that govern these reactions, allowing chemists to design more effective catalysts and optimize reaction conditions.

Materials Science:

Quantum simulations can predict the properties of new materials, including superconductors, batteries, and nanomaterials. By modeling the electronic structure of these materials, researchers can identify candidates with desirable properties for specific applications.

Applications in Cosmology

In the realm of cosmology, quantum simulations offer new avenues for exploring the universe's origins and fundamental structure.

Modeling the Early Universe:

Quantum simulations can help recreate conditions in the early universe, allowing researchers to investigate the behavior of matter and energy during critical phases of cosmic evolution. This modeling may provide insights into the Big Bang, cosmic inflation, and the formation of structures in the universe.

Understanding Dark Matter and Dark Energy:

Dark matter and dark energy remain among the biggest mysteries in cosmology. Quantum simulations can be used to model the interactions and properties of these elusive components, aiding in the quest to understand their roles in the universe's expansion and structure.

Quantum Gravity:

One of the greatest challenges in modern physics is reconciling quantum mechanics with general relativity. Quantum simulations can be instrumental in exploring theories of quantum gravity, potentially shedding light on phenomena like black holes and the fabric of spacetime itself.

Case Studies: Pioneering Quantum Simulations

Simulating the Hydrogen Molecule:

Researchers have successfully used quantum computers to simulate the simplest molecule, hydrogen, demonstrating the accuracy and efficiency of quantum simulations. This milestone showcases the potential for simulating more complex systems in the future.

Exploring Chemical Reactions:

In a recent study, scientists employed quantum simulations to model a series of chemical reactions, providing insights into reaction dynamics and intermediates. This research paves the way for more efficient drug discovery and materials design.

Studying Quantum Phase Transitions:

Researchers have simulated quantum phase transitions in spin systems, revealing unexpected behavior that could have implications for understanding topological phases of matter.

Challenges and Future Directions

Despite the exciting potential of quantum simulations, several challenges remain:

Error Correction:

Quantum computers are susceptible to errors due to decoherence and noise. Developing robust error correction techniques is essential for achieving reliable quantum simulations.

Scalability:

Building larger and more powerful quantum computers capable of simulating complex systems remains a significant challenge. Ongoing research aims to enhance the scalability of quantum hardware.

Interdisciplinary Collaboration:

Quantum simulations require expertise from various fields, including physics, chemistry, and computer science. Encouraging interdisciplinary collaboration will be crucial for maximizing the potential of quantum simulations.

CHAPTER 18:

Quantum Energy: Redefining Power Systems

As the world grapples with the urgent need for sustainable energy solutions, the quest for efficient, clean, and reliable energy sources has never been more critical. Quantum technologies, which leverage the principles of quantum mechanics, hold tremendous potential to revolutionize energy systems across various sectors. From quantum batteries that promise unprecedented energy storage capabilities to optimizing power grids and discovering new renewable energy sources, the intersection of quantum computing and energy presents a pathway to a sustainable future. This chapter will explore how quantum technologies can redefine power systems and address some of the most pressing energy challenges of our time.

The Energy Landscape: Current Challenges

The global energy landscape is characterized by several significant challenges that demand innovative solutions:

Energy Storage Limitations: Traditional batteries face limitations in terms of energy density, charge-discharge rates, and lifespan. As renewable energy sources like solar and wind become more prevalent, effective energy storage solutions are essential to balance supply and demand.

Grid Management and Efficiency: Modern power grids are often inefficient, suffering from energy losses during transmission and facing challenges in integrating variable renewable energy sources. As electricity demand continues to rise, optimizing grid management is crucial.

Decarbonization Efforts: The transition to renewable energy is essential for reducing carbon emissions and combating climate change. However, many existing energy systems are heavily reliant on fossil fuels, making the shift to sustainable sources challenging.

Resource Management: Managing the variability of renewable energy sources requires advanced forecasting and real-time data analysis to ensure a stable energy supply. Current systems often struggle to effectively manage these fluctuations.

Quantum Technologies: A New Frontier in Energy

Quantum technologies offer innovative solutions to these energy challenges, potentially transforming how we generate, store, and distribute power.

Quantum Batteries: One of the most exciting prospects in quantum energy is the development of quantum batteries—devices that use quantum mechanics to store and release energy more efficiently than classical batteries.

High Energy Density: Quantum batteries have the potential to offer significantly higher energy densities, allowing them to store more energy in a smaller footprint. This property is particularly valuable for applications such as electric vehicles and portable electronics.

Faster Charge-Discharge Cycles: Due to quantum effects like superposition and entanglement, quantum batteries may be capable of charging and discharging at unprecedented rates. This capability can reduce downtime and enhance the usability of energy storage systems.

Self-Discharging Control: Quantum batteries could offer advanced control over self-discharging processes, allowing users to maintain energy levels for extended periods without loss, making them more reliable for various applications.

Optimizing Power Grids: Quantum computing can enhance the management of power grids by optimizing the flow of electricity and improving efficiency.

Real-Time Data Processing: Quantum algorithms can analyze vast amounts of data from sensors throughout the grid, allowing for real-time monitoring and adjustments. This capability can enhance grid reliability and reduce energy losses.

Load Balancing: Quantum computing can optimize load distribution across the grid, ensuring that electricity demand is met without overloading any single component. This optimization can enhance grid resilience and stability.

Integration of Renewables: By predicting fluctuations in renewable energy generation, quantum algorithms can facilitate the integration of solar and wind power into the grid, ensuring a smooth transition between energy sources and reducing reliance on fossil fuels.

Unlocking New Renewable Energy Sources: Quantum technologies may also play a crucial role in discovering and optimizing new renewable energy sources.

Materials Discovery: Quantum computing can accelerate the discovery of new materials for solar cells, batteries, and fuel cells by simulating molecular interactions and properties. This capability can lead to more efficient and cost-effective energy technologies.

Optimizing Energy Conversion: Quantum simulations can help optimize processes like photosynthesis and catalysis, leading to more efficient methods of converting sunlight or biomass into usable energy.

Hydrogen Production: Quantum technologies may assist in developing more efficient methods for hydrogen production, which is considered a promising clean energy source. Understanding molecular interactions through quantum simulations can enhance catalytic processes used in hydrogen production.

Case Studies: Quantum Innovations in Energy

Quantum Battery Research: Researchers are exploring various quantum battery designs that utilize quantum superposition and entanglement. Early experiments have demonstrated the potential for rapid charge-discharge cycles and high energy densities, paving the way for commercial applications.

Smart Grid Optimization: Utility companies are beginning to leverage quantum computing to enhance grid management. Pilot projects have shown improved load balancing and integration of renewable energy sources, resulting in increased efficiency and reduced costs.

Material Science and Energy: Collaborations between materials scientists and quantum computing experts are underway to discover new materials for solar energy applications. Initial findings suggest that quantum simulations can expedite the identification of materials with optimal properties for energy conversion.

Ethical Considerations and Challenges

As quantum technologies are integrated into energy systems, several ethical considerations and challenges must be addressed:

Equitable Access to Technology: Ensuring that advancements in quantum energy are accessible to all communities, including underserved populations, is crucial. Policymakers and researchers must prioritize inclusivity in technology development.

Environmental Impact Assessment: The deployment of new energy technologies should be accompanied by comprehensive environmental assessments to evaluate potential impacts on ecosystems and communities.

Regulatory Frameworks: As quantum technologies disrupt existing energy systems, regulatory frameworks must be adapted to accommodate these innovations while ensuring safety, reliability, and fairness.

CHAPTER 19:

The Ethics of Quantum Computing

As quantum computing rapidly advances, it promises to bring unprecedented capabilities across numerous fields, from cryptography and drug discovery to artificial intelligence and climate modeling. However, with these advancements come significant ethical considerations that must be addressed to ensure that quantum technologies are developed and deployed responsibly. This chapter explores the key ethical challenges associated with quantum computing, including privacy, inequality, and the potential misuse of quantum technologies.

1. Privacy and Data Security

The advent of quantum computing poses profound implications for data privacy and security. Classical encryption methods, which currently protect sensitive information such as personal data, financial transactions, and government communications, are at risk due to the superior computational power of quantum computers.

Breaking Classical Encryption: Quantum algorithms, such as Shor's algorithm, can efficiently factor large numbers, rendering widely used encryption protocols like RSA vulnerable. This capability threatens the confidentiality of communications and sensitive data, raising concerns about privacy violations.

Quantum Key Distribution (QKD): While quantum computing may undermine classical encryption, it also offers solutions like Quantum Key Distribution (QKD), which uses the principles of quantum mechanics to create secure communication channels. However, the implementation of QKD is still in its infancy, and questions remain about its practicality and scalability for widespread use.

Data Ownership and Control: As quantum technologies advance, the issue of data ownership becomes increasingly complex. Individuals may lose control over their personal data as quantum systems enable new methods of data analysis and surveillance. Addressing who owns and controls data in a quantum context is crucial to protecting individual privacy.

2. Inequality and Access to Technology

The rapid development of quantum computing has the potential to exacerbate existing inequalities in society. Access to advanced technologies often correlates with socioeconomic status, educational background, and geographic location, creating disparities that could hinder equitable distribution of benefits.

Access to Resources: Quantum computing requires substantial investments in research, infrastructure, and education. Institutions and countries with greater financial resources may advance more quickly, leaving less affluent regions behind. This disparity could result in a digital divide, where only a select few benefit from quantum advancements.

Job Displacement: As quantum computing automates complex tasks and optimizes processes, there is a risk of job displacement in sectors reliant on traditional computing. Addressing potential job losses and ensuring that affected individuals have opportunities for retraining and reskilling is essential for fostering an inclusive transition to a quantum-driven economy.

Education and Workforce Development: Developing a skilled workforce to meet the demands of the quantum computing industry is crucial. Efforts must be made to ensure that educational resources and training opportunities are accessible to diverse populations, including underrepresented communities.

3. Potential Misuse of Quantum Technologies

With the power of quantum computing comes the potential for misuse in various contexts, raising ethical concerns about its applications in society.

Cyber Warfare: The ability of quantum computers to break classical encryption could lead to new forms of cyber warfare, where malicious actors exploit quantum technologies to compromise sensitive information and infrastructure. This possibility necessitates international cooperation to establish norms and protocols for responsible use.

Surveillance and Control: Quantum technologies may enable advanced surveillance capabilities, raising concerns about government overreach and violations of civil liberties. The potential for misuse in monitoring individuals and controlling populations must be carefully considered, ensuring that quantum advancements do not compromise democratic values.

Weaponization of Quantum Technologies: The development of quantum technologies for military applications raises ethical questions about the potential for new forms of warfare. As nations invest in quantum capabilities, establishing guidelines for the responsible development and use of these technologies is vital.

4. Environmental Considerations

As the quantum computing industry expands, its environmental impact cannot be overlooked. The manufacturing and operation of quantum computers require energy and resources, and the environmental footprint of these technologies must be evaluated.

Sustainable Practices: Developing quantum technologies in a manner that prioritizes sustainability is crucial. Efforts should be made to minimize energy consumption and environmental harm during production and operation.

Life Cycle Assessment: Conducting life cycle assessments of quantum technologies can provide insights into their environmental impact, guiding decisions toward more sustainable practices in research and development.

5. Ethical Frameworks and Governance

To address the ethical considerations of quantum computing, it is essential to establish frameworks and governance structures that promote responsible development and use.

Interdisciplinary Collaboration: Engaging ethicists, technologists, policymakers, and representatives from diverse communities is crucial to ensure that ethical considerations are integrated into the development of quantum technologies. Collaborative efforts can foster a holistic understanding of the potential implications.

Regulatory Oversight: As quantum technologies continue to advance, regulatory bodies must adapt to the unique challenges posed by quantum computing. Developing regulations that address privacy, security, and equitable access will be essential to protect individuals and society.

Public Awareness and Engagement: Promoting public awareness of quantum technologies and their implications is vital for informed discussions about ethical considerations. Engaging the public in conversations about the future of quantum computing can empower individuals to advocate for responsible practices.

CHAPTER 20:

Democratizing Quantum: Ensuring Global Access

As quantum computing continues to evolve at a rapid pace, the potential benefits it offers could reshape industries, drive innovations, and tackle some of the most pressing challenges facing humanity. However, these advantages also risk creating a "quantum divide," where access to quantum technologies becomes confined to wealthy individuals and nations, exacerbating existing inequalities and leaving underprivileged communities behind. This chapter explores the implications of this divide and presents actionable strategies to ensure global access to quantum computing for all.

1. The Quantum Divide: Risks and Implications

The emergence of quantum computing raises the specter of a significant disparity between those who can leverage its capabilities and those who cannot. Several factors contribute to this potential divide:

Resource Disparities: Quantum computing development requires substantial financial investments in research, infrastructure, and skilled personnel. Wealthy nations and well-funded private entities are better positioned to make these investments, potentially monopolizing the benefits of quantum technologies.

Education and Expertise: Access to education and training in quantum computing is often concentrated in elite institutions, limiting opportunities for individuals in lower-income regions. The resulting knowledge gap could prevent a significant portion of the global population from participating in the quantum revolution.

Innovation Concentration: As quantum technologies advance, innovative applications and breakthroughs may primarily originate from a few regions, perpetuating a cycle where developing countries struggle to keep pace, limiting their ability to address local challenges effectively.

2. The Importance of Global Inclusion

Ensuring global access to quantum computing is essential not only for equitable development but also for fostering innovation that benefits all of humanity. By democratizing quantum technology, we can:

Encourage Diverse Perspectives: A broader range of voices and experiences can lead to more creative and inclusive solutions to global challenges. Diverse participation in the quantum space will ensure that the technologies developed reflect a variety of needs and values.

Foster Collaborative Research: International collaboration in quantum research can lead to advancements that benefit everyone, from improving healthcare in developing nations to optimizing agricultural practices worldwide.

Mitigate Risks of Monopolization: Promoting global access helps to mitigate the risks associated with the monopolization of quantum technologies by a few entities, reducing the likelihood of negative social, economic, and political consequences.

3. Strategies for Ensuring Global Access

To prevent a quantum divide and promote global access to quantum computing, a multifaceted approach is needed, involving governments, academia, industry, and civil society. Here are some strategies that can be employed:

Investment in Education and Workforce Development: Governments and institutions should invest in educational initiatives that promote quantum literacy and provide training opportunities for individuals at all levels. This includes creating online courses, scholarships, and mentorship programs to reach underrepresented communities.

International Collaboration and Partnerships: Fostering partnerships between universities, research institutions, and industries across countries can facilitate knowledge sharing and resource pooling. Collaborative initiatives can drive innovation and ensure that breakthroughs benefit a wider audience.

Open-Source Quantum Software and Tools: Developing and promoting open-source quantum computing platforms and tools can democratize access to quantum resources. By making quantum software freely available, individuals and organizations can experiment with and develop their own applications, regardless of their financial resources.

Infrastructure Development: Investing in the infrastructure needed for quantum computing, such as cloud-based quantum services, can help bridge the gap for regions lacking physical access to quantum hardware. Cloud-based quantum computing allows users to access powerful quantum processors remotely, democratizing access to cutting-edge technology.

Policy Frameworks and Regulation: Governments should establish policies that encourage equitable access to quantum technologies. This includes promoting fair competition, protecting intellectual property rights, and ensuring that the benefits of quantum advancements are shared across society.

Engaging Non-Governmental Organizations (NGOs): NGOs can play a vital role in advocating for equitable access to quantum technologies, especially in developing nations. By raising awareness, providing resources, and facilitating capacity-building initiatives, NGOs can help empower communities to participate in the quantum revolution.

4. Case Studies: Successful Initiatives

Quantum Education Initiatives: Organizations such as the Quantum Education Foundation and the Quantum Open Source Foundation are working to create educational resources, workshops, and training programs aimed at building quantum literacy among diverse populations worldwide.

International Research Collaborations: Initiatives like the Quantum Internet Alliance and the Quantum Computing for Development project bring together researchers from different countries to collaborate on quantum projects that address global challenges.

Cloud-Based Quantum Services: Companies like IBM and Google offer cloud access to their quantum processors, allowing researchers and developers from around the world to experiment with quantum algorithms and applications without the need for significant investments in hardware.

5. The Role of Policy and Governance

To ensure the successful implementation of these strategies, robust policy frameworks and governance structures are essential:

Global Dialogue and Cooperation: International organizations, governments, and industry leaders must engage in dialogue to establish norms and principles for the ethical development and use of quantum technologies.

Support for Developing Countries: Wealthier nations should provide support and resources to help developing countries build their quantum computing capabilities, fostering an inclusive global quantum ecosystem.

Monitoring and Evaluation: Establishing metrics for assessing progress toward equitable access to quantum computing will help identify gaps and areas for improvement, ensuring that efforts remain focused on promoting inclusion.

A Collective Responsibility

Democratizing access to quantum computing is not merely a technical challenge; it is a moral imperative that requires collective action from all stakeholders. By addressing the risks of a quantum divide and implementing strategies for global inclusion, we can harness the transformative potential of quantum technologies for the benefit of all.

The journey toward democratizing quantum computing is a shared responsibility that calls for innovative solutions, collaborative efforts, and unwavering commitment to equity. As we navigate the complexities of this new frontier, we must prioritize inclusivity, ensuring that the advantages of quantum technologies are accessible to everyone, regardless of their background or circumstances.

In doing so, we can unlock the full potential of quantum computing to drive meaningful progress, tackle pressing global challenges, and create a future where innovation serves humanity as a whole, not just the privileged few. As we look to the horizon, let us strive for a quantum future that reflects our highest ideals and aspirations for a just and equitable world.

CHAPTER 21:

Quantum Governance: Policymaking in the Quantum Age

As quantum computing matures and begins to infiltrate various sectors, the need for robust governance frameworks becomes increasingly evident. The unique capabilities and potential risks associated with quantum technologies necessitate comprehensive regulatory approaches that not only address the potential for misuse but also foster innovation and collaboration across borders. This chapter explores the principles and practices of quantum governance, emphasizing the delicate balance between regulation and innovation.

1. The Need for Quantum Governance

The unprecedented power of quantum computing brings forth challenges that classical regulatory frameworks are ill-equipped to address. Several factors underscore the necessity for tailored governance in the quantum era:

Potential for Misuse: Quantum technologies could be misused for malicious purposes, including cyberattacks, surveillance, and the manipulation of data. Effective governance must establish safeguards against such threats.

Rapid Pace of Innovation: The fast-evolving nature of quantum computing requires adaptive regulatory frameworks that can keep pace with technological advancements. Rigid regulations could stifle innovation and hinder the growth of the quantum ecosystem.

Global Implications: The transnational nature of quantum technologies necessitates international cooperation and agreements to address shared challenges and promote responsible development.

2. Key Principles of Quantum Governance

To navigate the complexities of quantum governance, policymakers should adhere to several key principles:

Flexibility and Adaptability: Regulatory frameworks must be flexible enough to accommodate the rapid advancements in quantum technologies. Policymakers should adopt iterative approaches that allow for adjustments as the technology evolves.

Collaboration and Stakeholder Engagement: Involving a diverse range of stakeholders—including scientists, technologists, industry leaders, and civil society—in the policymaking process ensures that regulations are informed by various perspectives and expertise.

Ethical Considerations: Governance frameworks should prioritize ethical considerations, addressing issues such as privacy, equity, and potential societal impacts. Regulations should reflect societal values and protect the rights of individuals.

Transparency and Accountability: Policymakers must promote transparency in the development and deployment of quantum technologies. Establishing clear accountability mechanisms can help build public trust and ensure responsible use.

3. Regulatory Challenges and Considerations

Governments and international bodies face several challenges in establishing effective governance frameworks for quantum technologies:

Defining the Regulatory Scope: Policymakers must determine which aspects of quantum technologies require regulation, including quantum computing, quantum cryptography, and quantum communication. Balancing regulation with the need for innovation can be challenging.

Interdisciplinary Approaches: Quantum technologies intersect with various fields, including cybersecurity, telecommunications, and finance. Effective governance requires interdisciplinary collaboration to address the multifaceted implications of quantum advancements.

Addressing Global Disparities: Regulations must consider the diverse capabilities and resources of countries in developing and deploying quantum technologies. Ensuring that regulations do not disproportionately disadvantage developing nations is essential for fostering global equity.

4. Strategies for Effective Quantum Governance

Policymakers can adopt several strategies to establish effective governance frameworks for quantum technologies:

Establishing Regulatory Sandboxes: Creating regulatory sandboxes allows innovators to test new quantum applications and services in a controlled environment. These frameworks can facilitate experimentation while ensuring compliance with necessary regulations.

International Cooperation and Treaties: Governments should engage in international cooperation to establish treaties and agreements that address the global implications of quantum technologies. Collaborative efforts can help mitigate risks and promote responsible practices.

Creating Quantum Ethics Committees: Establishing ethics committees dedicated to quantum technologies can provide guidance on ethical considerations, ensuring that the development and deployment of these technologies align with societal values.

Developing Standards and Best Practices: Policymakers should work with industry stakeholders to develop standards and best practices for quantum technologies. Establishing benchmarks can enhance security, interoperability, and public trust.

5. Case Studies in Quantum Governance

Several initiatives and organizations have begun to explore quantum governance, serving as valuable examples of best practices:

The Quantum Industry Consortium: This organization fosters collaboration among quantum technology stakeholders, including industry, academia, and government. By developing common standards and frameworks, the consortium aims to drive responsible innovation in quantum technologies.

The European Union's Quantum Technologies Strategy: The EU has recognized the importance of quantum technologies and has established a strategy to promote research, innovation, and regulation. The strategy emphasizes ethical considerations, international collaboration, and public engagement.

The National Quantum Initiative in the United States: This initiative aims to accelerate quantum research and development while addressing the need for secure and ethical use of quantum technologies. The initiative promotes collaboration between government, academia, and industry to foster a responsible quantum ecosystem.

6. The Role of Public Engagement

Engaging the public in discussions about quantum technologies and their implications is essential for effective governance. Policymakers should prioritize public awareness campaigns and educational initiatives to foster understanding and encourage informed dialogue.

Building Public Trust: Transparency in policymaking and the involvement of the public can help build trust in quantum governance. Engaging diverse communities ensures that regulations reflect societal values and address the concerns of all stakeholders.

Encouraging Citizen Participation: Policymakers should create avenues for citizen participation in the governance process, allowing individuals to voice their concerns and contribute to discussions about the future of quantum technologies.

A New Paradigm for Governance

As we navigate the complexities of the quantum age, establishing effective governance frameworks for quantum technologies is essential for maximizing their potential while minimizing risks. Policymakers must embrace flexibility, collaboration, and ethical considerations as guiding principles in their efforts to regulate this transformative technology.

Quantum governance represents a new paradigm in policymaking —one that recognizes the interconnectedness of technology, society, and ethics. By fostering innovation and ensuring responsible use, we can unlock the full potential of quantum technologies to address global challenges and improve the human experience.

CHAPTER 22:

Quantum Warfare: The New Arms Race

As nations grapple with the transformative power of quantum computing, a new arena of competition emerges: quantum warfare. With the potential to revolutionize military strategies, enhance cybersecurity, and redefine the balance of power, quantum technologies are poised to become a critical component of modern warfare. This chapter delves into the implications of quantum computing for defense strategies, its potential offensive uses, and the global race for quantum dominance.

1. The Quantum Arms Race: An Introduction

The advent of quantum computing has prompted a re-evaluation of military capabilities worldwide. Countries are investing heavily in research and development to harness quantum technologies for national security purposes. This quantum arms race has implications for defense policies, military strategies, and international relations.

Historical Context: Just as the atomic bomb transformed warfare in the 20th century, quantum technologies are set to redefine military capabilities in the 21st century. Nations that successfully develop quantum capabilities may gain significant strategic advantages over their adversaries.

Emerging Competitors: Countries like the United States, China, and Russia are leading the charge in quantum research, with military implications at the forefront of their strategies. The race for quantum supremacy extends beyond technology to include global influence and leadership in defense.

2. Defense Strategies Enhanced by Quantum Technologies

Quantum computing has the potential to enhance various defense strategies, including:

Quantum Cryptography for Secure Communications: Quantum key distribution (QKD) offers unbreakable encryption based on the principles of quantum mechanics. This technology allows military communications to remain secure from interception, providing a significant advantage in sensitive operations.

Advanced Threat Detection and Response: Quantum computing's ability to process vast amounts of data in real time can enhance threat detection capabilities. By analyzing data from sensors and intelligence sources, quantum systems can identify potential threats and respond more quickly than traditional systems.

Autonomous Systems and Decision-Making: Quantum algorithms could improve the performance of autonomous systems, such as drones and robots, by enabling them to make real-time decisions based on complex datasets. This could enhance the effectiveness of military operations and reduce the risks to human personnel.

3. Potential Offensive Uses of Quantum Computing

While quantum technologies can enhance defensive capabilities, they also present new opportunities for offensive strategies:

Cyber Warfare: Quantum computing poses both a threat and an opportunity in cyber warfare. While it could break classical encryption methods, it can also be used to develop new offensive cyber tools that exploit vulnerabilities in enemy systems. Nations may seek to use quantum algorithms to launch sophisticated cyber attacks against critical infrastructure.

Weaponization of Quantum Technologies: The potential for quantum-enhanced weapons raises ethical and strategic concerns. For example, quantum radar systems could provide unprecedented tracking capabilities, making it difficult for adversaries to conceal their movements. Additionally, advancements in quantum simulations could lead to the development of new types of weapons or military strategies.

Information Warfare: The ability to manipulate and control information in the quantum realm could lead to new forms of psychological and information warfare. Nations may leverage quantum technologies to spread disinformation, disrupt communication networks, or create confusion in enemy ranks.

4. The Role of International Treaties and Agreements

As the quantum arms race intensifies, the need for international cooperation and regulatory frameworks becomes paramount. Key considerations include:

Preventing an Arms Race: Nations must work together to establish agreements that prevent the proliferation of quantum weapons and promote responsible use of quantum technologies in military contexts.

Transparency and Confidence-Building Measures: Open dialogues and transparency in quantum research can help build trust among nations, reducing the likelihood of misunderstandings and conflicts.

Ethical Guidelines for Quantum Warfare: Developing ethical guidelines for the use of quantum technologies in warfare is essential to address concerns about their potential impact on civilians and global security.

5. Case Studies in Quantum Warfare

Several nations are leading the way in integrating quantum technologies into their military strategies:

The United States: The U.S. Department of Defense has prioritized quantum research as part of its national security strategy. Investments in quantum computing and cryptography aim to enhance military capabilities and maintain technological superiority.

China: China has made significant advancements in quantum communication and cryptography, positioning itself as a leader in the field. The Chinese military is exploring the integration of quantum technologies into its defense systems.

Russia: Russia is investing in quantum technologies for military applications, focusing on enhancing cybersecurity and developing advanced weapons systems that leverage quantum principles.

6. The Future of Quantum Warfare

As quantum technologies continue to advance, the landscape of warfare will evolve in profound ways. The implications for military strategies, international relations, and global security will be significant:

The Balance of Power: Nations that successfully develop and integrate quantum capabilities into their military strategies may gain a decisive edge, reshaping the balance of power on the global stage.

Increased Competition and Tension: The race for quantum dominance could lead to heightened competition and tensions between nations, potentially resulting in conflicts fueled by technological rivalries.

Ethical Dilemmas: The use of quantum technologies in warfare raises ethical questions about the impact on civilians, the environment, and the nature of conflict itself. Policymakers must grapple with these dilemmas as they navigate the future of quantum warfare.

Navigating the Quantum Frontier

Quantum warfare represents a new frontier in military strategy, where the potential for transformative capabilities must be balanced against ethical considerations and the need for international cooperation. As nations race to develop and deploy quantum technologies, the importance of robust governance frameworks and ethical guidelines cannot be overstated.

By fostering transparency, collaboration, and responsible use of quantum technologies, we can mitigate the risks associated with quantum warfare and promote global stability. The future of warfare in the quantum age will depend not only on technological prowess but also on our ability to navigate the complex ethical and geopolitical landscape that accompanies these advancements.

CHAPTER 23:

Quantum Diplomacy: Global Cooperation

As quantum computing continues to evolve, its implications extend beyond national borders, presenting a unique opportunity—and necessity—for international collaboration. The complexity and potential impact of quantum technologies demand a cooperative approach that includes shared research initiatives, collective cybersecurity strategies, and the establishment of ethical guidelines for their use. This chapter explores the concept of quantum diplomacy, emphasizing the role of international organizations such as the United Nations (UN) and the North Atlantic Treaty Organization (NATO) in fostering global cooperation on quantum issues.

1. The Imperative for International Collaboration

Quantum technologies possess the power to reshape not only individual nations but also the global landscape. As such, collaborative efforts are crucial for several reasons:

Shared Risks and Threats: Quantum computing can both enhance national security and pose significant risks, particularly regarding cybersecurity. As nations develop quantum capabilities, the potential for cyber threats increases. Cooperative efforts can help mitigate these risks through shared knowledge and strategies.

Global Competitiveness: Countries that collaborate on quantum research can accelerate innovation and maintain a competitive edge in the global arena. Joint research initiatives can lead to breakthroughs that benefit all participating nations, fostering a sense of unity and shared purpose.

Addressing Ethical Dilemmas: The ethical considerations surrounding quantum technologies, such as their potential impact on privacy and security, necessitate a coordinated response. Establishing common ethical guidelines can help ensure responsible use of quantum capabilities across borders.

2. The Role of International Organizations

Organizations like the UN and NATO are positioned to play pivotal roles in facilitating quantum diplomacy. Their frameworks can help foster cooperation among nations in the following ways:

Creating Platforms for Dialogue: The UN can provide a neutral ground for discussions on quantum technologies, enabling member states to share concerns, best practices, and collaborative opportunities. By fostering dialogue, the UN can help build trust among nations and reduce the potential for conflict.

Establishing Regulatory Frameworks: The UN and other international bodies can work to establish regulatory frameworks that govern the use of quantum technologies. These frameworks can address issues such as cybersecurity standards, arms control, and ethical guidelines.

Coordinating Research Initiatives: International organizations can facilitate joint research initiatives, pooling resources and expertise to accelerate advancements in quantum technologies. By coordinating efforts, nations can share the financial burden of research while maximizing the potential for innovation.

3. Key Areas for Collaboration

Several key areas require international cooperation to effectively address the challenges posed by quantum technologies:

Quantum Cybersecurity: As quantum computing threatens traditional encryption methods, nations must collaborate on developing new quantum-resistant encryption techniques. Joint efforts in cybersecurity can enhance resilience against potential attacks and foster a sense of collective security.

Research and Development: Collaborative research initiatives can drive advancements in quantum computing, quantum communication, and quantum cryptography. Countries can pool resources to tackle significant scientific challenges, leading to breakthroughs that benefit all participating nations.

Ethical Guidelines and Standards: Establishing ethical guidelines for the development and deployment of quantum technologies is essential. International organizations can work together to create standards that prioritize ethical considerations, ensuring responsible use while addressing concerns about privacy, security, and equity.

4. Case Studies in Quantum Diplomacy

Several initiatives and partnerships illustrate the potential for quantum diplomacy in fostering global cooperation:

The European Quantum Flagship: The European Union has launched the Quantum Flagship initiative, which aims to coordinate quantum research across member states. This collaborative effort focuses on developing quantum technologies while promoting ethical considerations and public engagement.

The Global Quantum Communication Network: Several countries, including China, the United States, and European nations, have initiated projects to develop a global quantum communication network. This collaborative effort aims to create secure communication channels that leverage quantum key distribution.

The NATO Science for Peace and Security Program: NATO's Science for Peace and Security (SPS) program fosters collaboration among member states in science and technology, including quantum research. By promoting joint projects and information sharing, NATO can enhance collective security and technological advancement.

5. Challenges to Quantum Diplomacy

While the potential for collaboration is immense, several challenges must be addressed to realize effective quantum diplomacy:

National Security Concerns: Nations may be reluctant to share sensitive information related to quantum technologies due to national security concerns. Building trust and establishing clear frameworks for information sharing will be essential to overcoming these barriers.

Differing Priorities and Capacities: Countries have varying levels of expertise and resources when it comes to quantum research. Establishing equitable partnerships that recognize these differences will be crucial for successful collaboration.

Competing Geopolitical Interests: The geopolitical landscape can complicate collaborative efforts, as nations may prioritize their own interests over collective action. Promoting a shared vision for the benefits of quantum technologies will be essential in overcoming these challenges.

6. The Path Forward: Building a Quantum Diplomatic Framework

To effectively harness the potential of quantum technologies through diplomacy, several strategies can be employed:

Creating a Quantum Coalition: Nations can establish a coalition focused on quantum technologies, bringing together stakeholders from government, industry, and academia. This coalition can facilitate dialogue, collaboration, and knowledge sharing, driving progress in quantum research and applications.

Promoting Public-Private Partnerships: Engaging the private sector in quantum diplomacy can enhance research and innovation. By fostering public-private partnerships, governments can leverage industry expertise and resources to advance quantum technologies while ensuring alignment with ethical guidelines.

Fostering Education and Public Engagement: Raising awareness and understanding of quantum technologies among the public is crucial for building support for international cooperation. Education initiatives can help demystify quantum computing and its implications, fostering a more informed global citizenry.

A Unified Approach to Quantum Diplomacy

Quantum diplomacy represents a vital avenue for addressing the challenges and opportunities presented by quantum technologies. By fostering international cooperation through dialogue, research collaboration, and ethical guidelines, nations can collectively harness the potential of quantum computing for the greater good.

As we stand at the intersection of technology and diplomacy, it is essential to recognize that the future of quantum technologies hinges on our ability to work together. By prioritizing cooperation and understanding, we can navigate the complexities of the quantum era and ensure that the benefits of quantum advancements are shared equitably across the globe.

CHAPTER 24:

Quantum Computing in Space Exploration

The final frontier of space exploration is poised to undergo a revolutionary transformation through the capabilities offered by quantum computing. As humanity pushes the boundaries of our knowledge and exploration of the universe, quantum technologies will provide the tools needed to optimize spacecraft design, enhance communication systems, and analyze vast amounts of astronomical data. This chapter delves into the ways in which quantum computing will accelerate space exploration and the new frontiers it may unlock.

1. The Need for Quantum Solutions in Space Exploration

Space exploration is a complex endeavor that demands innovative solutions to overcome the myriad challenges faced beyond our atmosphere. Some of these challenges include:

Complex Problem Solving: The design and navigation of spacecraft involve numerous variables and constraints, often leading to highly complex optimization problems. Traditional computing methods may struggle with the computational demands required for such intricate calculations.

Communication Limitations: Spacecraft operate at vast distances from Earth, leading to latency and communication challenges. Current communication systems can be vulnerable to interference and loss of data, necessitating the development of more robust systems.

Data Overload: The vast amounts of data generated by astronomical observations and spacecraft sensors require advanced processing capabilities. Traditional computing systems may not be equipped to handle the scale and complexity of this data in real time.

2. Quantum Computing for Optimizing Spacecraft Design

One of the most promising applications of quantum computing in space exploration is optimizing spacecraft design. Quantum algorithms can help streamline various aspects of the design process:

Material Science and Engineering: Quantum simulations can enhance our understanding of material properties, leading to the development of lightweight, durable materials that can withstand the extreme conditions of space. By simulating the interactions of particles at the quantum level, researchers can discover new materials with revolutionary properties tailored for specific missions.

Trajectory Optimization: Quantum computing can be employed to solve complex trajectory optimization problems, determining the most efficient paths for spacecraft to minimize fuel consumption and travel time. This capability is particularly crucial for missions to distant planets or celestial bodies, where efficient navigation can save time and resources.

System Integration: The integration of various subsystems within a spacecraft requires balancing multiple design parameters. Quantum algorithms can assist in identifying optimal configurations that enhance performance and reliability while minimizing costs and risks.

3. Enhancing Communication Systems with Quantum Technology

Communication is critical in space exploration, especially for coordinating missions and ensuring data integrity. Quantum computing can enhance communication systems in several ways:

Quantum Key Distribution (QKD): Quantum cryptography offers unbreakable encryption through QKD, which can secure communications between spacecraft and ground stations. This ensures that sensitive mission data remains protected from potential cyber threats.

Quantum Communication Networks: Researchers are exploring the development of quantum communication networks that could enable instantaneous and secure data transmission over vast distances. These networks could revolutionize communication between spacecraft and Earth, facilitating real-time monitoring and control.

Interference Mitigation: Quantum technologies can enhance signal processing capabilities, allowing for improved reception and transmission of data in the face of interference. This is particularly valuable in environments with high levels of noise, such as those experienced during deep-space missions.

4. Data Analysis and Astrophysics: Harnessing Quantum Power

The field of astrophysics generates massive amounts of data, from telescope observations to data collected from space missions. Quantum computing can enhance data analysis in several key areas:

Accelerating Data Processing: Quantum algorithms can dramatically reduce the time required to analyze large datasets, enabling researchers to extract meaningful insights from astronomical observations much faster than classical computing methods.

Complex Simulations: Quantum simulations can model complex cosmic phenomena, such as black hole formations, galaxy collisions, and stellar evolution. By accurately simulating these processes, scientists can gain deeper insights into the universe's behavior and origins.

Machine Learning Applications: Quantum computing can enhance machine learning algorithms, improving pattern recognition in astronomical data. This can lead to better identification of celestial objects, anomaly detection, and the discovery of new astronomical phenomena.

5. Case Studies: Quantum Initiatives in Space Exploration

Several initiatives are underway to explore the potential of quantum computing in space exploration:

NASA's Quantum Computing Research: NASA is actively investigating the applications of quantum computing for space missions. Research projects focus on optimization problems related to spacecraft design and mission planning, as well as quantum algorithms for analyzing astronomical data.

The European Space Agency (ESA): The ESA has initiated projects exploring quantum communication technologies for secure data transmission between satellites and ground stations. Their efforts aim to enhance the resilience and security of satellite communications.

University Collaborations: Various universities and research institutions are partnering with space agencies to explore the integration of quantum technologies into space exploration. Collaborative efforts focus on developing quantum algorithms for trajectory optimization and enhancing communication systems.

6. The Future of Quantum Computing in Space Exploration

The integration of quantum computing into space exploration presents exciting possibilities for the future:

Interstellar Missions: As humanity looks toward interstellar exploration, quantum computing could play a crucial role in optimizing mission trajectories and communication over vast distances. Quantum technologies may enable more feasible missions to distant stars and galaxies.

Enhanced Scientific Discovery: The ability to process and analyze astronomical data at unprecedented speeds will accelerate scientific discovery. Researchers can quickly identify and study new celestial objects, leading to a deeper understanding of the universe.

International Collaborations: The challenges of space exploration will likely necessitate global collaboration, and quantum computing can serve as a unifying force. Joint research initiatives and shared technologies can foster international partnerships in space exploration, ultimately benefiting humanity as a whole.

The Quantum Frontier in Space Exploration

As we venture into the cosmos, quantum computing offers unprecedented opportunities to accelerate space exploration and enhance our understanding of the universe. From optimizing spacecraft design to revolutionizing communication systems, quantum technologies will become indispensable tools for scientists and engineers.

The potential of quantum computing extends far beyond what we can imagine today, opening doors to new discoveries and advancements in space exploration. By harnessing the power of quantum technologies, we can navigate the complexities of the universe, tackle the challenges of interstellar travel, and deepen our understanding of the cosmos.

CHAPTER 25:

Redefining the Economy in the Quantum Era

As quantum computing begins to permeate various sectors of society, its transformative effects on the global economy are becoming increasingly apparent. The introduction of quantum technologies is poised to redefine traditional economic structures, influencing everything from labor markets to productivity levels and innovation-driven economies. This chapter explores how the quantum revolution will shape economic landscapes, create new opportunities, and pose unique challenges in a rapidly evolving world.

1. The Quantum Advantage: Unlocking New Economic Potential

Quantum computing represents a significant leap forward in computational capabilities, which will enable businesses and industries to perform tasks that were previously thought impossible. The potential benefits of quantum computing for the economy include:

Increased Productivity: Quantum computers can process information at speeds exponentially faster than classical computers. This capability will lead to increased productivity across industries, as complex calculations and data analysis can be performed in a fraction of the time.

Enhanced Decision-Making: The ability to analyze vast datasets quickly will improve decision-making processes in business and government. Companies can leverage quantum algorithms to optimize operations, forecast market trends, and enhance supply chain efficiency.

Innovation Acceleration: Quantum computing will facilitate breakthroughs in various fields, including pharmaceuticals, materials science, and energy. This acceleration of innovation will lead to the creation of new markets, products, and services, ultimately driving economic growth.

2. Transforming Labor Markets

The introduction of quantum computing will have profound implications for labor markets, reshaping job roles and required skill sets:

Shifting Job Demands: As quantum technologies become more integrated into business operations, the demand for workers skilled in quantum computing and related fields will rise. This shift may lead to a labor market that increasingly values expertise in quantum algorithms, data science, and cybersecurity.

Job Displacement and Creation: While some traditional jobs may be displaced due to automation and efficiency gains, the quantum economy will also create new opportunities in emerging sectors. Preparing the workforce for this transition will require a focus on reskilling and upskilling initiatives.

Education and Training: To meet the demands of a quantum-driven economy, educational institutions must adapt their curricula to include quantum computing fundamentals. Collaboration between universities, businesses, and government will be essential to develop training programs that equip individuals with the skills needed for the future workforce.

3. Quantum Computing and Economic Inequality

The adoption of quantum technologies may exacerbate existing economic inequalities if access is not managed equitably:

The Quantum Divide: As quantum computing capabilities develop, a "quantum divide" may emerge between nations and regions with access to advanced technologies and those without. This divide could hinder economic growth in less developed areas, perpetuating disparities in wealth and opportunity.

Ensuring Inclusive Access: Addressing the quantum divide will require proactive measures to ensure equitable access to quantum computing resources. Governments, educational institutions, and private enterprises must collaborate to create pathways for underserved communities to participate in the quantum economy.

4. Global Competition and Innovation-Driven Economies

The race to develop quantum technologies is shaping global competition and influencing national economic strategies:

Strategic Investments: Countries around the world are making strategic investments in quantum research and development to maintain competitiveness. Nations that prioritize quantum initiatives may gain a significant advantage in emerging industries and technological leadership.

Innovation Ecosystems: The quantum era will foster the creation of innovation ecosystems that connect startups, research institutions, and established companies. These ecosystems will drive collaboration and accelerate the development of quantum technologies, resulting in economic growth and job creation.

International Collaboration: Collaboration among nations in quantum research can enhance global competitiveness. Sharing knowledge and resources can lead to breakthroughs that benefit multiple countries and contribute to shared economic prosperity.

5. Case Studies: Quantum Innovations and Economic Impact

Several initiatives illustrate the transformative economic impact of quantum computing:

IBM Quantum Experience: IBM's Quantum Experience platform democratizes access to quantum computing resources, enabling researchers, businesses, and educators to experiment with quantum algorithms. This initiative fosters innovation and collaboration, driving economic growth in quantum-related fields.

Google Quantum AI: Google's Quantum AI team focuses on developing quantum algorithms to solve real-world problems in industries such as finance and logistics. Their work is expected to create economic opportunities by optimizing processes and reducing costs.

D-Wave Systems: D-Wave, a leader in quantum computing, offers quantum annealing technology that companies can use to tackle complex optimization problems. This technology has applications across various industries, from logistics to finance, illustrating the potential for economic transformation.

6. Preparing for the Quantum Economy

To navigate the shift toward a quantum-driven economy, several strategies can be implemented:

Policy Development: Governments must develop policies that encourage quantum innovation while addressing potential challenges, such as workforce displacement and economic inequality. These policies should promote research funding, education initiatives, and equitable access to quantum resources.

Fostering Public-Private Partnerships: Collaboration between public and private sectors is essential to drive quantum research and development. By leveraging resources and expertise, these partnerships can accelerate advancements and ensure that the benefits of quantum technologies are widely shared.

Encouraging Lifelong Learning: In a rapidly changing economic landscape, promoting a culture of lifelong learning will be critical. Individuals must continuously update their skills to remain competitive in the quantum economy, and educational institutions should offer accessible training programs.

Embracing the Quantum Future

The transformative impact of quantum computing on the global economy presents both opportunities and challenges. As we stand on the cusp of the quantum era, it is essential to embrace innovation, collaboration, and inclusivity to redefine economic structures for the betterment of society.

By harnessing the power of quantum technologies, we can unlock new levels of productivity, enhance decision-making, and accelerate the pace of innovation. The future of work will require a proactive approach to workforce development and education, ensuring that individuals are equipped to thrive in the evolving quantum landscape.

CHAPTER 26:

The Quantum Workforce: Jobs of the Future

As quantum computing technology evolves, it is set to transform the job market in profound ways. The rise of quantum computing will not only create new job opportunities but also reshape existing roles, necessitating a significant shift in the skills required for the workforce of the future. This chapter explores the implications of quantum computing on employment, the essential skills for upcoming generations, and the educational reforms necessary to prepare the workforce for a quantum-driven economy.

1. The Emerging Landscape of Quantum Jobs

Quantum computing will create a plethora of new job roles across various sectors, from technology and finance to healthcare and research. Some key areas of job growth include:

Quantum Software Development: As more businesses adopt quantum computing, the demand for software developers skilled in quantum algorithms and programming languages will surge. Roles such as quantum software engineers, quantum algorithm developers, and quantum systems architects will become essential.

Quantum Research Scientists: Researchers focused on advancing quantum technologies will be in high demand. These scientists will work on developing new quantum algorithms, improving quantum hardware, and exploring innovative applications of quantum computing across disciplines.

Quantum Data Analysts: With the capacity to analyze vast amounts of data, quantum computing will give rise to new positions focused on data analysis. Quantum data analysts will leverage quantum algorithms to derive insights from complex datasets, significantly improving decision-making processes.

Quantum Cybersecurity Experts: As quantum computing challenges existing encryption methods, the need for cybersecurity professionals skilled in quantum-safe cryptography will grow. These experts will design security protocols to protect sensitive information in the quantum era.

Interdisciplinary Roles: Quantum computing's impact will extend beyond traditional tech roles. Fields such as finance, healthcare, and logistics will require professionals who understand both quantum principles and their applications in their respective domains. For instance, financial analysts who can utilize quantum computing to optimize investment strategies or healthcare professionals who can harness quantum technologies for personalized medicine.

2. Skills Required for the Quantum Economy

The quantum workforce will necessitate a new set of skills that go beyond traditional education. Key competencies will include:

Quantum Computing Fundamentals: A strong understanding of quantum mechanics, qubits, superposition, entanglement, and quantum gates will be essential for professionals in the field. This foundational knowledge will enable individuals to grasp the principles of quantum computing and apply them effectively.

Programming Skills: Familiarity with quantum programming languages such as Qiskit, Cirq, and Quipper will be critical. Professionals will need to develop proficiency in coding for quantum systems, creating algorithms that leverage quantum advantages.

Data Science and Machine Learning: As quantum computing enhances data analysis capabilities, skills in data science and machine learning will be increasingly valuable. Professionals who can integrate quantum algorithms with machine learning techniques will be at the forefront of innovation.

Interdisciplinary Knowledge: Given the diverse applications of quantum computing, professionals will benefit from interdisciplinary knowledge that bridges quantum computing with fields such as finance, biology, and logistics. This approach will foster collaboration and enable the development of innovative solutions.

Critical Thinking and Problem Solving: The ability to approach complex problems creatively and analytically will remain a core competency. As quantum technologies tackle challenges previously deemed insurmountable, professionals will need to think critically and devise effective solutions.

3. Educational Changes for the Quantum Future

Preparing future generations for the quantum workforce will require significant changes in educational systems at all levels. Key initiatives may include:

Curriculum Integration: Educational institutions must integrate quantum computing concepts into existing STEM curricula. Introducing quantum mechanics, quantum algorithms, and programming languages at an early age will help students build a strong foundation.

Interdisciplinary Programs: Universities should develop interdisciplinary programs that combine quantum computing with other fields, such as finance, healthcare, and materials science. These programs will prepare students to tackle real-world challenges using quantum technologies.

Hands-On Learning and Research Opportunities: Providing students with hands-on experience in quantum computing is crucial. Initiatives such as internships, research projects, and partnerships with quantum companies can offer valuable opportunities for practical learning.

Online Courses and Workshops: Online platforms can facilitate access to quantum computing education, making it easier for individuals to upskill or reskill in this emerging field. Offering courses and workshops focused on quantum programming and applications will broaden the talent pool.

Collaboration with Industry: Educational institutions should collaborate with industry partners to align curricula with the skills needed in the workforce. Involving professionals in curriculum development ensures that students acquire relevant skills for quantum-related careers.

4. The Role of Government and Industry

Both government and industry play crucial roles in shaping the future quantum workforce:

Policy Support: Governments can promote education and workforce development initiatives by providing funding for quantum research, scholarships, and training programs. Policies that support collaboration between academia and industry will also enhance workforce readiness.

Industry Partnerships: Companies in the quantum sector can partner with educational institutions to develop training programs and internships. By working together, businesses can ensure that graduates possess the necessary skills and knowledge for the evolving job market.

Continuous Learning and Professional Development: Encouraging a culture of continuous learning will be essential as quantum technologies evolve. Companies can offer ongoing training and professional development opportunities to help employees stay current in this rapidly changing field.

5. Case Studies: Successful Quantum Workforce Initiatives

Several organizations and educational institutions are already paving the way for the quantum workforce:

IBM Quantum Education: IBM has launched various initiatives to educate the next generation of quantum computing professionals. Their Quantum Educators Program provides resources for educators and students, while the IBM Quantum Challenge encourages hands-on learning through competitions.

Microsoft Quantum Program: Microsoft offers a comprehensive quantum computing curriculum through its Quantum Development Kit. The program provides resources for learning quantum programming and algorithms, supporting the development of a skilled quantum workforce.

University Partnerships: Universities such as MIT, Caltech, and the University of Waterloo are establishing quantum computing research centers and offering specialized degrees in quantum information science. These initiatives prepare students for careers in the quantum economy and foster collaboration between academia and industry.

6. Conclusion: Embracing the Quantum Workforce Transformation

The advent of quantum computing heralds a significant transformation in the job market, requiring new skills and educational frameworks to prepare future generations. As quantum technologies reshape industries, the demand for a skilled workforce will grow, offering exciting opportunities for individuals willing to adapt and learn.

By investing in education, fostering interdisciplinary collaboration, and promoting continuous learning, society can prepare for the quantum workforce of the future. Embracing the challenges and opportunities presented by quantum computing will not only drive economic growth but also position individuals and communities to thrive in an increasingly complex and interconnected world.

CHAPTER 27:

Quantum Education: Redefining Learning

As quantum computing begins to reshape various industries and society at large, the need for a new approach to education becomes increasingly urgent. Quantum literacy is emerging as a fundamental skill set necessary for the workforce of the future, alongside traditional STEM (Science, Technology, Engineering, and Mathematics) skills. This chapter explores the imperative for educational reform, the development of innovative curricula, and the adoption of effective teaching methods that will prepare students for a world increasingly influenced by quantum technologies.

1. The Imperative for Quantum Literacy

Quantum literacy encompasses an understanding of quantum principles and the ability to apply them in practical contexts. The rising importance of quantum literacy can be attributed to several factors:

Technological Advancement: Quantum technologies are progressing rapidly, with implications for industries such as computing, healthcare, finance, and materials science. A workforce equipped with quantum literacy will be better positioned to innovate and adapt in this changing landscape.

Competitive Advantage: Nations and organizations that prioritize quantum education will have a competitive edge in the global economy. Cultivating a pool of quantum-literate individuals will drive research, development, and economic growth.

Informed Citizenry: As quantum technologies become more prevalent, it is crucial for the general public to understand their implications. Quantum literacy will empower individuals to engage in informed discussions about the ethical, societal, and policy issues related to quantum computing.

2. Curriculum Development: Meeting the Needs of the Quantum Era

To foster quantum literacy, educational institutions must develop curricula that integrate quantum concepts across various subjects. Key components of an effective quantum education curriculum include:

Interdisciplinary Approach: Quantum concepts should be integrated into existing STEM curricula as well as the humanities and social sciences. By showing the intersections of quantum mechanics with economics, ethics, and art, educators can provide students with a holistic understanding of quantum's impact on society.

Practical Applications: Curricula should emphasize the real-world applications of quantum computing. Case studies, simulations, and hands-on projects can illustrate how quantum technologies are being used in fields such as medicine, finance, and environmental science.

Modular Learning: Educational programs should be designed with flexibility in mind, allowing students to explore quantum topics at their own pace. Modular learning structures can accommodate diverse learning styles and enable students to deepen their understanding of specific quantum concepts.

Emphasis on Critical Thinking: Educators should prioritize developing students' critical thinking and problem-solving skills. Engaging students in inquiry-based learning and encouraging them to tackle complex challenges will prepare them to navigate the intricacies of quantum technologies.

3. Innovative Teaching Methods for Quantum Education

In addition to developing new curricula, educational institutions must adopt innovative teaching methods to engage students in quantum learning. Effective strategies include:

Project-Based Learning: Hands-on projects that require students to apply quantum principles to solve real-world problems can enhance understanding and retention. Collaborative projects can also foster teamwork and communication skills.

Flipped Classrooms: In a flipped classroom model, students engage with learning materials outside of class (e.g., through videos and readings) and use class time for discussions, problem-solving, and hands-on activities. This approach allows educators to spend more time addressing student questions and exploring complex quantum topics in depth.

Virtual and Augmented Reality: Integrating virtual and augmented reality technologies can provide immersive experiences that deepen students' understanding of quantum concepts. These technologies can help visualize abstract quantum phenomena and make complex topics more accessible.

Interdisciplinary Collaborations: Partnerships between different departments can enhance quantum education. For example, combining physics and computer science courses can provide students with a comprehensive understanding of both quantum mechanics and quantum computing.

4. Preparing Educators for Quantum Teaching

For effective quantum education to take root, educators themselves must be well-versed in quantum concepts. Key strategies for preparing educators include:

Professional Development: Ongoing professional development opportunities should be available for educators to learn about quantum computing and related technologies. Workshops, online courses, and partnerships with quantum research institutions can help teachers stay updated.

Collaboration with Quantum Experts: Building partnerships with quantum researchers and professionals can enrich the educational experience. Guest lectures, collaborative projects, and mentorship opportunities can expose educators and students to real-world applications of quantum technologies.

Resource Development: Creating teaching resources, including textbooks, online materials, and interactive simulations, will help educators effectively teach quantum concepts. Open-access resources can enhance availability and accessibility.

5. Case Studies: Institutions Leading the Way

Several institutions are pioneering innovative quantum education initiatives:

The University of Waterloo's Institute for Quantum Computing: The University of Waterloo offers specialized programs and courses focused on quantum information science. Their interdisciplinary approach combines physics, engineering, and computer science, preparing students for careers in quantum technologies.

MIT's Quantum Education Initiative: MIT has developed a Quantum Education Initiative that includes undergraduate courses, research opportunities, and community engagement activities. Their curriculum emphasizes hands-on learning and collaboration with industry partners.

Quantum Computing for Everyone: Numerous online platforms, such as edX and Coursera, offer courses on quantum computing, making it accessible to a broader audience. These platforms empower individuals to learn at their own pace and explore quantum topics without the constraints of traditional education.

6. The Role of Government and Industry in Quantum Education

To support quantum education, government and industry must collaborate in several key areas:

Funding and Grants: Governments can provide funding for quantum education initiatives, supporting research, curriculum development, and teacher training programs.

Industry Partnerships: Collaboration between educational institutions and quantum technology companies can enhance curriculum relevance and provide students with practical experiences through internships and research projects.

Public Awareness Campaigns: Raising awareness about the importance of quantum literacy can encourage interest in quantum education. Public campaigns highlighting the relevance of quantum computing to various fields can inspire students to pursue careers in this area.

Embracing a Quantum Future through Education

The rise of quantum computing necessitates a fundamental rethinking of education to equip future generations with the skills and knowledge needed to thrive in a quantum-driven world. By fostering quantum literacy and integrating quantum concepts into curricula, educational institutions can prepare students for the challenges and opportunities that lie ahead.

Innovative teaching methods and interdisciplinary approaches will engage students and cultivate critical thinking skills essential for navigating the complexities of quantum technologies. By investing in educator training and collaboration with industry, we can ensure a skilled workforce ready to embrace the quantum future.

CHAPTER 28:

Quantum Computing in Entertainment

The entertainment industry is on the cusp of a technological revolution, with quantum computing poised to transform how content is created, delivered, and experienced. From gaming to film and interactive storytelling, quantum technologies will enable unprecedented advancements that redefine the boundaries of creativity and immersion. This chapter explores the ways in which quantum computing will revolutionize entertainment, enhancing realism, improving production efficiencies, and inspiring new forms of storytelling.

1. The Current State of Entertainment Technology

Before diving into the transformative impact of quantum computing, it's essential to understand the current landscape of entertainment technology:

Classical Computing Limitations: Traditional computing has enabled significant advancements in graphics, sound design, and interactive experiences. However, it often faces limitations in processing power, particularly when rendering complex simulations or managing large datasets.

Realism in Content Creation: While advancements in graphics engines, motion capture, and visual effects have led to increasingly realistic experiences, achieving true-to-life simulation of environments, characters, and interactions remains a challenge.

Data Management: The explosion of user-generated content and the growing demand for high-quality streaming services require robust data management systems. Classical systems struggle to efficiently process and analyze the vast amounts of data generated by both creators and consumers.

2. Enhancing Realism with Quantum Simulations

Quantum computing offers unique capabilities that can significantly enhance realism in entertainment. Key areas of impact include:

Advanced Visual Effects: Quantum simulations can model complex physical interactions, such as fluid dynamics and light behavior, in real time. This capability allows filmmakers and game developers to create highly detailed and realistic visual effects that were previously impossible or too resource-intensive to achieve.

Character Behavior and Animation: Quantum computing can facilitate more sophisticated algorithms for character behavior and animation. By simulating a broader range of physical and emotional states, developers can create characters that respond more naturally to in-game situations, enhancing immersion and engagement.

Virtual Worlds and Environments: Quantum technologies can model intricate virtual worlds with rich details, enabling creators to design expansive environments that react dynamically to user interactions. This results in more engaging and lifelike experiences for players and viewers alike.

3. Transforming Game Development and Design

The gaming industry is set to benefit immensely from the integration of quantum computing. Potential advancements include:

Real-Time Processing Power: Quantum computers can handle vast datasets and complex algorithms simultaneously, allowing for real-time processing of game environments, player interactions, and artificial intelligence (AI) behaviors. This capability enables richer and more responsive gameplay experiences.

Dynamic Content Generation: Quantum algorithms can be employed to generate game content dynamically based on player choices, behaviors, and preferences. This leads to personalized gaming experiences that adapt to individual players, keeping them engaged longer.

Improved AI Capabilities: Quantum computing can significantly enhance AI in gaming, allowing for more complex decision-making processes and more intelligent NPC (non-player character) behavior. This can create a more challenging and engaging gaming environment.

4. Revolutionizing Film Production

Quantum computing has the potential to streamline and enhance various aspects of film production:

Speeding Up Render Times: Rendering high-quality visual effects is one of the most time-consuming aspects of film production. Quantum computing can drastically reduce render times, allowing filmmakers to iterate on their designs more rapidly and effectively.

Enhanced Post-Production: Quantum technologies can improve post-production processes, such as color grading and audio mixing. By using quantum algorithms to analyze and optimize audio and visual elements, filmmakers can achieve higher quality results in less time.

Collaborative Production: Quantum computing could facilitate more efficient collaboration across teams and locations. By leveraging quantum communication technologies, multiple production teams can work on complex simulations and visual effects simultaneously, increasing overall productivity.

5. Interactive Storytelling: The New Frontier

Quantum computing opens up exciting possibilities for interactive storytelling, allowing creators to craft experiences that adapt and evolve based on user input:

Branching Narratives: With quantum algorithms, storytellers can develop branching narratives that change based on viewers' decisions. This approach enables more complex storylines that feel personalized to each viewer.

Real-Time Story Adjustments: Quantum computing allows for real-time adjustments to narratives based on audience reactions and engagement metrics. This means that a story can evolve dynamically as viewers interact with it, creating a unique experience for each audience member.

Immersive Experiences: By combining virtual reality (VR) and quantum computing, creators can design immersive environments where users influence the story and interact with characters in meaningful ways. This form of storytelling blurs the lines between creator and audience, allowing for a collaborative narrative experience.

6. Case Studies: Quantum Innovations in Entertainment

As the integration of quantum computing in entertainment is still in its infancy, there are promising developments worth highlighting:

Quantum Game Development Initiatives: Companies like D-Wave are exploring how quantum computing can be utilized for game development. By developing prototypes that leverage quantum algorithms, they aim to create new gaming experiences that are richer and more engaging.

Film Production Optimization: Film studios are beginning to explore partnerships with quantum computing firms to optimize rendering processes and enhance visual effects. By utilizing quantum simulations, studios hope to speed up production timelines and improve final outputs.

Interactive Media Experiences: Projects that focus on interactive storytelling are being developed using quantum technologies. These projects aim to create narratives that adapt in real time, providing audiences with personalized and immersive experiences.

7. Challenges and Considerations

While the potential of quantum computing in entertainment is vast, there are challenges that must be addressed:

Access and Affordability: The current high costs of quantum computing hardware may limit access for smaller studios and independent creators. Ensuring that these technologies are affordable and accessible will be crucial for widespread adoption.

Skill Development: The integration of quantum computing into content creation requires a workforce skilled in both quantum mechanics and creative processes. Educational initiatives will be necessary to develop this talent pool.

Ethical Considerations: As with any technological advancement, ethical considerations surrounding the use of quantum computing in entertainment must be addressed. This includes concerns about data privacy, content ownership, and the potential for misuse of advanced simulation technologies.

A New Era of Entertainment Awaits

Quantum computing is set to revolutionize the entertainment industry, paving the way for more realistic simulations, dynamic content creation, and interactive storytelling experiences. By harnessing the power of quantum technologies, creators can push the boundaries of imagination and innovation, delivering engaging experiences that captivate audiences like never before.

CHAPTER 29:

The Arts in the Quantum Age

As we stand on the precipice of a new era defined by quantum computing, the realm of artistic expression is poised for unprecedented transformation. Quantum technologies have the potential to reshape not only how art is created but also how it is experienced and understood. This chapter delves into the intersection of quantum computing and the arts, exploring how artists and technologists can collaborate to generate innovative forms of digital art, immersive experiences, and new methods of creative expression.

1. The Evolution of Artistic Tools

Throughout history, advances in technology have significantly influenced artistic practices, from the invention of the camera to the advent of digital media. Quantum computing represents the next leap forward, providing artists with powerful tools to expand their creative horizons:

Enhanced Computational Power: Quantum computers can process vast amounts of data simultaneously, enabling artists to create intricate designs, simulations, and animations that would be impossible with classical computing. This power allows for more complex and nuanced artistic expressions.

New Algorithmic Possibilities: Artists can leverage quantum algorithms to generate unique patterns, textures, and forms, leading to novel artistic styles and aesthetics. The principles of superposition and entanglement can inspire new ways of thinking about composition and interaction.

Data-Driven Art: The ability to analyze large datasets in real time opens up new possibilities for data-driven art, where artists can create works that respond dynamically to information from the world around them. This can include everything from environmental data to social media trends, resulting in art that is both relevant and timely.

2. Quantum-Aided Creativity

Quantum computing can influence the creative process itself, offering new avenues for inspiration and exploration:

Generative Art: By employing quantum algorithms, artists can create generative art that evolves and changes based on specific parameters or interactions. This form of art challenges traditional notions of authorship and permanence, inviting audiences to engage with dynamic and ever-changing works.

Collaborative Creativity: Quantum technologies can facilitate collaboration between artists and scientists, merging artistic vision with scientific insight. Projects that blend these disciplines can yield innovative outcomes, leading to the creation of art that reflects complex scientific concepts.

Interactive Experiences: Quantum computing can enhance interactivity in artistic experiences, allowing audiences to engage with works in meaningful ways. Through real-time data analysis and quantum simulations, artworks can respond to viewers' actions, creating a dialogue between the observer and the observed.

3. Immersive Experiences and Virtual Reality

Quantum computing's potential to create immersive environments will revolutionize how audiences experience art:

Virtual Reality (VR) and Augmented Reality (AR): Quantum technologies can power advanced VR and AR applications, enabling artists to create fully immersive worlds that transport audiences to new dimensions. These environments can incorporate complex simulations that reflect the unpredictability and complexity of the quantum realm.

Multisensory Experiences: The integration of quantum computing with sensory technologies can create multisensory art experiences, where visual, auditory, and tactile elements combine to evoke profound emotional responses. By manipulating these elements in real time, artists can craft experiences that resonate deeply with audiences.

Quantum-Themed Installations: Artists can explore the themes and principles of quantum mechanics in their work, creating installations that invite audiences to contemplate the mysteries of the universe. These installations can blend art with science, sparking curiosity and dialogue about the nature of reality.

4. Case Studies: Artistic Innovations Inspired by Quantum Computing

Several pioneering artists and organizations are already exploring the potential of quantum computing in the arts:

Refik Anadol: A media artist known for his data-driven installations, Refik Anadol explores how algorithms can transform data into immersive experiences. By integrating quantum computing principles into his work, he pushes the boundaries of digital art, creating pieces that challenge perceptions of time and space.

Quantum Art Projects: Various interdisciplinary projects are emerging, where artists collaborate with quantum physicists to create works that reflect quantum principles. These projects often result in installations that visualize complex scientific concepts, making them accessible and engaging to a broader audience.

Generative Art and Music: Artists and musicians are beginning to experiment with quantum algorithms to create generative art and music. By using quantum computing to generate unique compositions in real time, they explore the intersections of technology, creativity, and randomness.

5. Challenges and Ethical Considerations

As quantum computing becomes more integrated into the arts, several challenges and ethical considerations arise:

Access and Equity: The high costs associated with quantum technologies may limit access for many artists, particularly those in marginalized communities. Ensuring equitable access to these tools will be crucial for fostering diverse voices in the quantum arts.

Ownership and Copyright: The collaborative nature of quantum-generated art raises questions about authorship and intellectual property. As generative art becomes more prevalent, new frameworks for understanding ownership and rights may be necessary.

Preservation of Digital Art: With the rapid pace of technological change, preserving digital art created using quantum computing will pose challenges. As quantum technologies evolve, ensuring the longevity and accessibility of these works will be essential for future generations.

Embracing a Quantum Future in the Arts

The integration of quantum computing into the arts represents a profound shift in how creativity is expressed and experienced. By harnessing the power of quantum technologies, artists can break free from traditional constraints and explore new dimensions of expression.

As we embrace this quantum future, it is essential to foster collaboration between artists, technologists, and scientists, creating a vibrant ecosystem that encourages innovation and experimentation. By doing so, we can unlock new forms of artistic expression that resonate with the complexities of the quantum world, inspiring audiences to engage with art in ways that challenge their perceptions and expand their understanding of reality.

CHAPTER 30:

Quantum in Sports and Fitness

In recent years, the intersection of technology and sports has led to groundbreaking advancements, with data analytics playing a crucial role in enhancing athletic performance. As quantum computing continues to develop, its transformative potential in sports and fitness becomes increasingly apparent. This chapter explores how quantum computing can revolutionize sports science by providing deeper insights into athlete performance, optimizing training programs, and ultimately pushing the boundaries of human capability.

1. The Current Landscape of Sports Science

Before exploring the potential of quantum computing, it's essential to understand the current state of sports science and data analytics:

Data-Driven Decisions: Traditional sports science heavily relies on data collection and analysis to inform decisions related to training, nutrition, and performance. Techniques such as motion capture, biometric sensors, and GPS tracking provide valuable insights into athletes' physical and psychological states.

Limitations of Classical Computing: While classical computers have enabled significant advancements in data analysis, they often struggle with the complexity of real-time processing, particularly when handling vast amounts of data from multiple sources. This limitation can impede the ability to derive actionable insights quickly.

The Role of Analytics: Advanced analytics have become integral to modern sports, allowing teams to evaluate player performance, develop game strategies, and make informed decisions about training regimens. However, the need for even more nuanced analysis remains.

2. The Power of Quantum Computing in Sports Analytics

Quantum computing has the potential to address many limitations of classical computing, enabling more sophisticated analysis of athletic performance:

Processing Large Datasets: Quantum computers can process enormous datasets at unprecedented speeds. This capability allows for the analysis of complex data generated by various sources—such as wearables, performance metrics, and environmental factors—simultaneously, providing a more comprehensive view of an athlete's performance.

Enhanced Predictive Modeling: Quantum algorithms can improve predictive modeling, allowing sports scientists to anticipate how different training regimens, recovery strategies, and even environmental conditions affect performance. This capability empowers coaches and athletes to tailor their approaches for optimal results.

Real-Time Insights: The speed of quantum computing enables real-time analysis of performance data during training and competitions. Coaches can access immediate feedback, allowing them to make quick adjustments to strategies or training protocols based on real-time metrics.

3. Optimizing Training Programs

Quantum computing can revolutionize how athletes train by optimizing programs to maximize performance while minimizing injury risk:

Personalized Training Regimens: By analyzing an athlete's performance data, recovery patterns, and physiological responses, quantum algorithms can develop personalized training programs that adapt to an individual's unique needs. This tailored approach can enhance performance and reduce the risk of overtraining or injury.

Biomechanical Analysis: Quantum computing can improve the analysis of biomechanics, allowing for a detailed understanding of an athlete's movement patterns. By simulating various techniques and providing insights into optimal biomechanics, coaches can refine athletes' techniques for improved performance.

Injury Prediction and Prevention: Using quantum analytics, sports scientists can develop models to predict injury risk based on training loads, biomechanics, and historical data. This predictive capability allows for proactive interventions, helping athletes avoid injuries before they occur.

4. Enhancing Performance Monitoring

Real-time performance monitoring is crucial for athletes and teams striving for excellence. Quantum computing can enhance this aspect in several ways:

Wearable Technology: The integration of quantum computing with wearable technology can lead to more accurate and sophisticated performance monitoring. Sensors that collect biometric data can be analyzed using quantum algorithms to provide insights into an athlete's fatigue levels, hydration status, and overall well-being.

Game Analysis and Strategy: Quantum computing can analyze game footage and player performance data, enabling teams to identify patterns and trends that inform strategies. This deep analysis can lead to more effective game plans and better in-game adjustments.

Team Dynamics and Cohesion: By analyzing interpersonal dynamics and communication patterns within teams, quantum computing can help coaches understand how team chemistry impacts performance. This insight can guide team-building strategies and enhance collaboration among athletes.

5. Case Studies: Quantum Innovations in Sports

As quantum computing is still in its early stages, various pilot projects are exploring its potential in sports:

Project Quantum Sports: This initiative involves collaborations between quantum computing companies and sports organizations to develop algorithms aimed at enhancing athlete performance. By combining quantum analytics with data from wearables, the project seeks to create a comprehensive performance monitoring system.

Athlete Performance Optimization: Some research institutions are conducting studies on the use of quantum algorithms for optimizing athlete training regimens. These studies aim to demonstrate how quantum computing can provide actionable insights that lead to improved performance outcomes.

Real-Time Game Analysis: Innovations in real-time data analysis for game strategies are being developed using quantum computing. By analyzing data from live games, teams can adapt their strategies on the fly, maximizing their chances of success.

6. Challenges and Considerations

While the potential of quantum computing in sports is vast, several challenges must be addressed:

Access to Technology: The high costs associated with quantum computing hardware and expertise may limit access for many sports organizations, particularly at the grassroots level. Ensuring equitable access will be crucial for fostering innovation across all levels of sport.

Data Privacy: As the use of data in sports analytics increases, concerns about data privacy and security will need to be addressed. Safeguarding athletes' personal data is essential for maintaining trust in data-driven approaches.

Integration with Existing Systems: Integrating quantum computing with existing sports science frameworks and practices may present challenges. Teams will need to develop strategies for effectively incorporating quantum insights into their training and performance protocols.

The Future of Sports in a Quantum Age

Quantum computing holds the promise of revolutionizing the field of sports science, enhancing athlete performance through advanced data analysis, optimizing training programs, and providing real-time insights. As technology continues to evolve, the potential for quantum innovations in sports will only grow.

By embracing quantum computing, athletes, coaches, and sports organizations can unlock new dimensions of performance, pushing the boundaries of what is possible in human athleticism. This evolution will require collaboration between technologists, scientists, and sports professionals to harness the full potential of quantum technologies in the pursuit of excellence.

CHAPTER 31:

Redefining Relationships: Quantum Psychology

As quantum computing continues to advance, its potential applications extend far beyond traditional fields like cryptography and material science. One of the most intriguing areas for exploration lies at the intersection of quantum computing and psychology. This chapter examines how quantum principles may enhance our understanding of the human mind, cognition, and behavior, ultimately redefining our approach to mental health and interpersonal relationships.

1. The Current Understanding of Psychology and Neuroscience

To appreciate the potential impact of quantum computing on psychology, it's essential to first understand the current state of psychological and neuroscientific research:

Classical Models of Cognition: Traditional psychology often relies on classical models that emphasize linear processing of information. These models help us understand behavior, decision-making, and cognitive biases, but they may not fully capture the complexity of human thought and emotion.

Neuroscientific Insights: Neuroscience has made significant strides in mapping the brain's structure and understanding how neural pathways influence behavior. However, the brain's intricate dynamics and the interplay between various cognitive processes remain only partially understood.

Limitations of Classical Approaches: Classical computing approaches to data analysis in psychology often struggle to manage the complexities and non-linearities inherent in human behavior. This limitation hampers the development of truly personalized therapeutic interventions.

2. The Quantum Approach to Understanding Cognition

Quantum computing introduces principles that challenge classical notions of processing and decision-making, offering new insights into cognitive science:

Superposition and Mental States: In quantum mechanics, superposition refers to a system being in multiple states simultaneously. Similarly, human cognition may be better understood as a complex interplay of various mental states. This concept can help us explore how individuals navigate ambiguous situations, weigh conflicting desires, and manage multiple identities.

Entanglement and Relationships: Quantum entanglement describes a phenomenon where particles become interconnected, such that the state of one immediately influences the state of another, regardless of distance. This idea can be applied to human relationships, suggesting that emotional bonds and interpersonal dynamics are more interconnected and complex than traditional psychology acknowledges.

Non-Linear Decision-Making: Quantum computing allows for non-linear decision-making processes, where multiple pathways can be explored simultaneously. This principle can shed light on how people make decisions under uncertainty, considering various potential outcomes rather than following a straightforward path.

3. Quantum Algorithms and Psychological Insights

The application of quantum algorithms to psychology could lead to significant advancements in understanding and analyzing cognitive processes:

Complex Data Analysis: Quantum computing excels at processing large datasets with intricate relationships. By applying quantum algorithms to psychological data, researchers can uncover hidden patterns and correlations that classical methods may overlook. This could lead to a deeper understanding of mental health conditions, personality traits, and cognitive styles.

Modeling Human Behavior: Using quantum models, psychologists can create simulations that more accurately reflect the complexities of human behavior. These simulations could account for multiple variables and interactions, allowing for more nuanced predictions about behavior in various contexts.

Optimizing Therapeutic Interventions: Quantum computing could enable the development of personalized therapeutic interventions. By analyzing individual cognitive patterns, emotional responses, and behavioral tendencies, mental health professionals could tailor treatments to address specific needs, leading to more effective outcomes.

4. Applications in Mental Health and Therapy

The integration of quantum principles into psychology could revolutionize mental health care:

Quantum-Enhanced Therapy: Therapies that incorporate insights from quantum psychology may provide more effective strategies for addressing mental health challenges. For example, understanding the interplay between various cognitive states can lead to more adaptive coping mechanisms for anxiety or depression.

Real-Time Data Monitoring: By leveraging quantum computing, mental health professionals could develop systems to monitor patients' emotional and cognitive states in real time. This capability would allow for immediate interventions and support, improving treatment outcomes.

New Approaches to Relationship Counseling: Understanding the entangled nature of human relationships could inform more effective approaches to relationship counseling. By recognizing the interconnectedness of emotions and behaviors, therapists could develop strategies that address underlying dynamics rather than simply addressing surface-level issues.

5. Case Studies: Exploring Quantum Psychology

While the field of quantum psychology is still in its infancy, early research and projects are beginning to explore its potential:

Interdisciplinary Research Initiatives: Universities and research institutions are forming interdisciplinary teams of psychologists, neuroscientists, and quantum physicists to explore the implications of quantum principles for understanding cognition and behavior. These initiatives aim to develop new theoretical frameworks and empirical studies.

Quantum-Informed Therapeutic Models: Some mental health practitioners are beginning to incorporate quantum principles into their therapeutic models, exploring how understanding cognitive complexity and interconnectedness can inform their practice.

Collaborative Projects: Collaborative projects involving technology companies and mental health organizations are emerging, focusing on developing quantum computing applications for data analysis in psychological research. These projects aim to create tools that enhance researchers' ability to analyze complex datasets.

6. Challenges and Ethical Considerations

As the field of quantum psychology evolves, several challenges and ethical considerations must be addressed:

Interpretation of Results: The interpretation of findings from quantum-informed psychological research may require a paradigm shift in how we understand cognition and behavior. Researchers must be cautious in how they communicate these insights to avoid misinterpretation.

Data Privacy: The collection and analysis of psychological data raise significant privacy concerns. Safeguarding individuals' personal information is essential to maintain trust in therapeutic processes and research initiatives.

Equity of Access: As quantum technologies develop, ensuring equitable access to these advancements in psychology will be crucial. Efforts must be made to prevent a divide between those who can afford cutting-edge treatments and those who cannot.

A New Era in Understanding the Mind

Quantum psychology offers a promising avenue for advancing our understanding of the human mind, cognition, and behavior. By integrating principles of quantum mechanics into psychological research and practice, we can explore the complexities of mental states and interpersonal relationships in new and profound ways.

As researchers, mental health professionals, and technologists collaborate to harness the power of quantum computing, we stand on the brink of a revolution in psychological science. This journey will not only enhance our understanding of the mind but also lead to more effective interventions, ultimately improving mental health outcomes for individuals and fostering healthier relationships.

CHAPTER 32:

Quantum Societies: How Communities Will Evolve

As quantum computing continues to develop and permeate various aspects of our lives, its influence on social structures and global interconnectedness is poised to reshape communities and societies as we know them. This chapter explores how the principles of quantum computing can impact human interactions, governance, and societal evolution, ultimately fostering a new era of collaboration and innovation.

1. The Foundation of Quantum Societies

To understand the emergence of quantum societies, we must first examine the characteristics of our current social structures:

Traditional Social Constructs: Traditional societies are often characterized by rigid hierarchies, centralized governance, and limited collaboration. Information flow is typically linear, with decisions made at higher levels before reaching the grassroots.

Digital Age Dynamics: The advent of the digital age has begun to disrupt these traditional structures, promoting greater connectivity and accessibility of information. However, significant barriers to global collaboration and equitable access remain, creating divides that hinder collective progress.

Globalization and Interconnectedness: As societies become more interconnected through globalization, the need for cooperative problem-solving and shared resources becomes increasingly important. Quantum computing has the potential to facilitate this interconnectedness on a larger scale.

2. The Role of Quantum Computing in Social Structures

Quantum computing introduces unique principles that could lead to the evolution of social structures and communities:

Decentralization and Collaboration: Quantum computing systems can operate in decentralized environments, promoting collaboration and shared decision-making. This shift can empower individuals and communities, enabling them to contribute meaningfully to societal challenges.

Enhanced Information Sharing: The ability to process vast amounts of data quickly allows for improved information sharing across communities. Real-time access to information fosters transparency and accountability, creating a more informed citizenry.

Collective Problem-Solving: Quantum algorithms can tackle complex societal issues that require collective intelligence. By leveraging diverse perspectives and expertise, communities can work together to address challenges like climate change, poverty, and public health.

3. Governance in Quantum Societies

As quantum computing reshapes social structures, governance will also evolve to reflect these changes:

Participatory Governance Models: Quantum societies may embrace participatory governance models that prioritize citizen engagement. Technology will enable more direct involvement in decision-making processes, allowing for a more democratic approach to governance.

Real-Time Policy Analysis: Quantum computing can facilitate real-time analysis of policy impacts, enabling governments to respond quickly to societal needs. This agility will enhance the ability to adapt policies based on data-driven insights, improving overall governance effectiveness.

Global Governance and Cooperation: As global challenges become more complex, quantum computing can enable international collaboration on governance. Countries can share data, resources, and expertise, working collectively to address issues that transcend borders.

4. Quantum Communities: Redefining Interactions

The principles of quantum computing have implications for how communities interact and collaborate:

Community Resilience: Quantum societies will be more resilient due to their collaborative nature. By leveraging collective intelligence, communities can adapt to changes and recover from disruptions more effectively.

Localized Solutions: Quantum computing can empower communities to develop localized solutions to global challenges. By analyzing data relevant to specific contexts, communities can design interventions that cater to their unique needs.

Cultural Exchange and Understanding: Enhanced connectivity through quantum technologies can foster cultural exchange and understanding among diverse communities. This interconnectedness promotes empathy and collaboration across cultural boundaries.

5. Case Studies: Early Innovations in Quantum Communities

Although the concept of quantum societies is still emerging, early initiatives demonstrate its potential:

Quantum Community Networks: Pilot projects are exploring the establishment of quantum community networks that enable decentralized decision-making and resource sharing. These networks empower citizens to collaborate on local issues, creating a more engaged and resilient society.

Collaborative Policy Platforms: Some governments are experimenting with collaborative policy platforms powered by quantum computing, allowing citizens to provide real-time feedback on proposed policies. This approach enhances transparency and fosters public trust.

Quantum-Enhanced Education Initiatives: Educational institutions are leveraging quantum computing to develop programs that promote global collaboration among students. These initiatives emphasize cross-cultural understanding and collective problem-solving, preparing future generations for a quantum society.

6. Challenges and Ethical Considerations

As we move toward quantum societies, several challenges and ethical considerations must be addressed:

Equitable Access to Technology: Ensuring equitable access to quantum computing technologies is essential to prevent exacerbating existing social inequalities. Efforts must be made to ensure that all communities can benefit from advancements in quantum computing.

Data Privacy and Security: The enhanced connectivity and data sharing inherent in quantum societies raise concerns about data privacy and security. Safeguarding personal information and establishing ethical data governance practices will be critical.

Navigating Disruption: The transition to quantum societies may lead to disruptions in traditional social structures and power dynamics. It is essential to manage these transitions thoughtfully to minimize resistance and promote inclusivity.

A New Era of Community and Collaboration

Quantum computing has the potential to redefine relationships within and between communities, fostering collaboration and interconnectedness. By breaking down traditional hierarchies and enabling real-time data analysis, quantum societies can address complex global challenges and create a more equitable and sustainable future.

As we embrace the possibilities of quantum computing, it is essential to prioritize ethical considerations, ensuring that all communities have access to the tools and technologies needed to thrive in a quantum society. By doing so, we can build a future where collaboration and innovation flourish, empowering individuals to contribute meaningfully to their communities and the world at large.

CHAPTER 33:

The Quantum City: Urban Planning and Smart Cities

As the world faces increasing urbanization, the demand for efficient and sustainable urban environments becomes paramount. Quantum technologies hold immense potential to transform our cities into smart ecosystems, optimizing everything from transportation systems to resource management and energy efficiency. This chapter explores how the principles of quantum computing can revolutionize urban planning, leading to the emergence of Quantum Cities that are more intelligent, interconnected, and sustainable.

1. The Rise of Smart Cities

Before delving into the impact of quantum technologies, it's essential to understand the current landscape of urban planning and the concept of smart cities:

Defining Smart Cities: Smart cities leverage technology to enhance the quality of life for their residents. This involves integrating information and communication technologies (ICT) to improve urban services, streamline operations, and promote sustainability.

Current Challenges: Despite the advancements in smart city initiatives, many urban areas still face significant challenges, including traffic congestion, energy waste, and inefficient resource allocation. These issues require innovative solutions that can adapt to the complexities of urban environments.

The Role of Data: Smart cities rely on data collected from various sensors, devices, and systems to inform decision-making and optimize services. However, the current data-processing methods can be limited, often unable to handle the sheer volume and complexity of data generated in urban settings.

2. Quantum Technologies: A Game Changer for Urban Planning

Quantum computing introduces a range of capabilities that can significantly enhance urban planning and management:

Complex Data Analysis: Quantum computers can process and analyze vast datasets far more efficiently than classical computers. This capability enables city planners to draw insights from data related to transportation patterns, energy consumption, and citizen behavior, leading to more informed decision-making.

Optimization Algorithms: Quantum algorithms excel at solving optimization problems, which are abundant in urban planning. For example, they can optimize traffic flow, resource allocation, and energy distribution, reducing waste and enhancing overall efficiency.

Predictive Modeling: Quantum technologies can improve predictive modeling for urban development, allowing planners to simulate the impacts of various scenarios, such as population growth, climate change, or infrastructural changes. This capability enables proactive planning rather than reactive responses.

3. Applications of Quantum Technologies in Urban Planning

Quantum technologies can transform multiple facets of urban planning, leading to smarter, more sustainable cities:

Intelligent Transportation Systems: Quantum computing can optimize traffic management by analyzing real-time data from various sources, including sensors, cameras, and GPS systems. This optimization can reduce congestion, improve public transportation efficiency, and enhance overall mobility in cities.

Energy Management: Smart cities can leverage quantum technologies to optimize energy distribution and consumption. By predicting demand patterns and analyzing renewable energy sources, cities can enhance energy efficiency and reduce carbon footprints.

Resource Management: Quantum algorithms can enhance resource management in areas such as water supply, waste management, and public services. By analyzing data in real time, cities can respond quickly to issues, allocate resources more effectively, and ensure that services are delivered efficiently.

4. Case Studies: Quantum Technologies in Action

Several pioneering cities and initiatives are beginning to explore the integration of quantum technologies into urban planning:

Quantum-Enabled Traffic Management: Some cities are piloting quantum algorithms for traffic management systems, analyzing historical data and real-time inputs to optimize traffic signals, reduce wait times, and improve overall flow.

Smart Grid Development: In regions focusing on renewable energy, quantum computing is being used to develop smart grid technologies that can predict energy demand fluctuations and optimize energy distribution from renewable sources.

Urban Resilience Planning: Cities facing the impacts of climate change are exploring quantum computing to develop predictive models for disaster response and recovery, enhancing their resilience to natural disasters and extreme weather events.

5. The Future of Urban Living: Quantum Cities

The integration of quantum technologies into urban planning leads to the conceptualization of Quantum Cities—urban environments that are:

Interconnected: Quantum cities will feature highly interconnected systems that allow for seamless communication between different services and infrastructure. This connectivity enhances overall efficiency and responsiveness to citizen needs.

Sustainable: By optimizing resource consumption and reducing waste, quantum cities will promote sustainability and resilience against climate challenges. They will leverage renewable energy sources and minimize environmental impact.

Inclusive: Quantum cities will prioritize inclusivity, using data-driven insights to ensure that all citizens have access to essential services and resources. Smart technologies will help identify and address disparities in urban living conditions.

6. Challenges and Ethical Considerations

While the potential of quantum technologies in urban planning is vast, several challenges and ethical considerations must be addressed:

Equitable Access to Technology: Ensuring that all communities have access to quantum technologies is essential to prevent exacerbating existing inequalities. Policymakers must prioritize inclusivity in the deployment of smart city initiatives.

Data Privacy and Security: The extensive data collection required for smart city management raises concerns about privacy and security. Robust frameworks must be established to protect citizens' data while ensuring transparency and accountability in its use.

Managing Technological Dependence: As cities become more reliant on quantum technologies, it's crucial to maintain a balance between technological dependence and human oversight. Cities must ensure that human decision-making remains central to urban governance.

Paving the Way for Quantum Cities

The integration of quantum technologies into urban planning holds the promise of creating intelligent, sustainable, and inclusive cities that can adapt to the complexities of modern life. By optimizing transportation, energy management, and resource allocation, Quantum Cities can enhance the quality of life for residents while addressing pressing global challenges.

As we move toward this new urban paradigm, it is vital to prioritize ethical considerations, ensuring equitable access to quantum technologies and safeguarding citizens' privacy. By fostering collaboration between governments, researchers, and communities, we can pave the way for the realization of Quantum Cities—urban environments that are not only technologically advanced but also responsive to the needs and aspirations of their residents.

CHAPTER 34:

Quantum Governance in Smart Cities

As cities around the world increasingly adopt smart technologies, the need for effective governance becomes paramount. Quantum computing, with its ability to process vast amounts of data and solve complex problems at unprecedented speeds, offers a transformative approach to urban governance. This chapter explores how quantum technologies can enable new forms of governance in smart cities, enhancing decision-making, crisis management, and citizen engagement.

1. The Evolution of Urban Governance

To understand the role of quantum computing in governance, it is essential to first examine the current landscape of urban governance:

Traditional Governance Models: Traditionally, urban governance has operated within hierarchical structures, where decision-making is often slow and reactive. Information flow is typically top-down, with limited input from citizens, leading to potential disconnects between government actions and community needs.

The Rise of Smart Cities: Smart city initiatives seek to integrate technology into urban governance, using data to improve service delivery, enhance quality of life, and promote sustainability. However, these initiatives often face challenges related to data management, integration, and real-time decision-making.

Need for Dynamic Governance: As cities become more complex and interconnected, there is a pressing need for governance models that are adaptive, transparent, and responsive to the needs of citizens. Quantum computing can play a crucial role in enabling this dynamic approach.

2. Quantum Technologies and Decision-Making

Quantum computing introduces several key capabilities that can revolutionize urban governance:

Data Processing Power: Quantum computers can analyze large datasets in real time, uncovering patterns and insights that classical computers struggle to identify. This capability allows city officials to make informed decisions based on accurate and comprehensive data.

Complex Simulations: Quantum technologies can simulate urban scenarios, helping policymakers understand the potential impacts of different decisions. By modeling various outcomes, city leaders can evaluate policies before implementation, reducing the risk of unintended consequences.

Predictive Analytics: Leveraging quantum algorithms, cities can enhance their predictive analytics capabilities. This means better forecasting of events such as traffic congestion, energy demand spikes, or public health crises, allowing for proactive measures and resource allocation.

3. Crisis Management through Quantum Governance

In the context of smart cities, crisis management is a critical area where quantum computing can provide significant advantages:

Real-Time Data Analysis: During emergencies, cities generate vast amounts of data from various sources, including sensors, social media, and public reports. Quantum computing can analyze this data in real time, enabling rapid response and informed decision-making during crises.

Resource Optimization: Quantum algorithms can help optimize resource allocation during emergencies, ensuring that essential services such as healthcare, transportation, and emergency response are deployed effectively. This optimization minimizes response times and enhances overall resilience.

Scenario Planning: Quantum computing allows for advanced scenario planning, enabling city officials to simulate different crisis situations and develop strategic responses. By evaluating various strategies in advance, cities can enhance their preparedness and adaptability.

4. Enhancing Citizen Engagement and Participation

Quantum governance also promotes greater citizen engagement in the decision-making process:

Transparent Decision-Making: By leveraging quantum computing to analyze and visualize data, city governments can present information transparently to citizens. This transparency fosters trust and allows residents to understand the rationale behind policy decisions.

Participatory Platforms: Quantum technologies can support participatory governance models, where citizens have direct input in decision-making. Online platforms powered by quantum computing can analyze public feedback and preferences, helping officials craft policies that reflect community needs.

Real-Time Feedback Mechanisms: Quantum governance can facilitate real-time feedback mechanisms, allowing citizens to provide input on services and initiatives as they happen. This responsiveness enhances accountability and encourages active citizen involvement.

5. Case Studies: Quantum Governance in Action

Emerging initiatives are beginning to illustrate the potential of quantum governance in smart cities:

Quantum-Enhanced Emergency Response Systems: Some cities are piloting quantum-based emergency response systems that leverage real-time data from sensors and citizen reports. By analyzing this data, these systems can identify hotspots for intervention and allocate resources effectively during crises.

Urban Planning Simulations: Urban planners are using quantum computing to simulate the impacts of new developments or policy changes, allowing for data-driven decisions that consider various community factors. This approach reduces conflicts and ensures that developments align with community needs.

Citizen-Centric Platforms: Innovative governance platforms are being developed that utilize quantum computing to process citizen feedback and preferences, enabling cities to prioritize initiatives based on real-time public sentiment.

6. Challenges and Ethical Considerations

While the potential for quantum governance in smart cities is immense, several challenges and ethical considerations must be addressed:

Data Privacy and Security: The use of vast amounts of data raises significant privacy and security concerns. Robust safeguards must be established to protect citizens' data and ensure that it is used ethically and transparently.

Digital Divide: Ensuring equitable access to quantum governance tools is essential to prevent widening the digital divide. Policymakers must prioritize inclusivity to ensure that all citizens can participate in and benefit from smart city initiatives.

Balancing Technology and Human Judgment: While quantum technologies offer powerful tools for decision-making, it is crucial to maintain a balance between technological insights and human judgment. Policymakers must remain at the forefront of governance, ensuring that technology complements rather than replaces human decision-making.

The Future of Quantum Governance

Quantum governance represents a transformative approach to managing smart cities, enabling dynamic, data-driven decision-making that enhances resilience, efficiency, and citizen engagement. By harnessing the power of quantum computing, city leaders can navigate the complexities of urban governance with greater agility and insight.

As we move toward the realization of quantum governance, it is vital to prioritize ethical considerations and ensure that all citizens can access and benefit from these advancements. By fostering collaboration between governments, technology providers, and communities, we can create a future where quantum governance not only enhances urban living but also empowers citizens to actively participate in shaping their cities.

CHAPTER 35:

Quantum Communication and the Media

In an age defined by rapid technological advancements and an ever-evolving media landscape, the advent of quantum communication is poised to revolutionize journalism, media dissemination, and information consumption. By leveraging the principles of quantum mechanics, such as entanglement and superposition, quantum communication offers unprecedented levels of security, efficiency, and speed, fundamentally transforming the way information flows through society. This chapter explores the potential impact of quantum communication on the media industry, the challenges it presents, and the opportunities for innovation in the dissemination of news and information.

1. The Current State of Media and Communication

To appreciate the revolutionary potential of quantum communication, it's essential to understand the current dynamics of media and communication:

Information Overload: With the explosion of digital media, consumers are inundated with information from various sources.

The sheer volume of content can lead to confusion, misinformation, and difficulties in discerning credible news.

Challenges of Security and Privacy: As journalism increasingly relies on digital platforms, issues related to data privacy and security have become prominent. High-profile data breaches and cyberattacks on media organizations raise concerns about the integrity of information and the safety of journalists and sources.

Speed vs. Accuracy: The demand for real-time reporting has pressured media organizations to prioritize speed over accuracy, sometimes resulting in the dissemination of misleading information. This phenomenon has led to a growing skepticism among consumers regarding the credibility of news sources.

2. Understanding Quantum Communication

Quantum communication utilizes the principles of quantum mechanics to transmit information securely and efficiently:

Quantum Key Distribution (QKD): One of the most significant advancements in quantum communication is quantum key distribution, which allows two parties to share a secret key for encrypting and decrypting messages. QKD is based on the fundamental principles of quantum mechanics, ensuring that any attempt to intercept the key would be detectable.

Quantum Entanglement: In quantum communication, entangled particles can be used to transmit information instantaneously over long distances, potentially enabling faster communication than classical systems. This principle can revolutionize the way news and information are shared across the globe.

Enhanced Data Security: Quantum communication offers a level of security that is theoretically invulnerable to hacking, as any attempt to eavesdrop on a quantum channel alters the state of the quantum system, alerting the communicating parties to potential breaches.

3. Transforming Journalism and Information Dissemination

Quantum communication presents several transformative opportunities for journalism and media:

Secure Reporting: Journalists often work with sensitive information, including whistleblower accounts and confidential sources. Quantum communication can provide a secure channel for transmitting sensitive information, protecting the identities of sources and the integrity of investigative reporting.

Real-Time News Delivery: The speed of quantum communication can facilitate real-time news delivery, allowing journalists to report breaking news instantly without the latency associated with traditional communication methods. This capability could enhance the immediacy and relevance of news coverage.

Authenticity Verification: Quantum communication can help authenticate news sources and verify the integrity of information. By utilizing quantum signatures, media organizations can establish the provenance of information, reducing the risk of misinformation and enhancing public trust.

4. The Impact on Media Consumption

The integration of quantum communication into media consumption will also change how audiences engage with news and information:

Personalized Content Delivery: Quantum algorithms can analyze user preferences and behaviors to deliver personalized content in real time. This capability allows media organizations to tailor news feeds and recommendations to individual users, enhancing engagement.

Enhanced Interactivity: Quantum communication can enable more interactive and immersive news experiences, allowing users to explore stories through augmented reality (AR) and virtual reality (VR) interfaces. This engagement can create deeper connections between audiences and the news.

Crowdsourced Journalism: With the security offered by quantum communication, crowdsourced journalism could become more prevalent. Citizen journalists can report news in real-time while ensuring their safety and confidentiality, contributing to a more diverse and inclusive media landscape.

5. Challenges and Ethical Considerations

While quantum communication offers numerous advantages, several challenges and ethical considerations must be addressed:

Implementation Costs: The infrastructure required for quantum communication can be expensive, posing challenges for media organizations, particularly smaller ones. Policymakers and industry leaders must explore ways to democratize access to quantum technologies.

Public Understanding and Acceptance: As with any emerging technology, public understanding and acceptance of quantum communication will be crucial. Educational initiatives may be necessary to inform consumers about the benefits and implications of quantum technologies in media.

Balancing Security and Transparency: While quantum communication enhances security, it is essential to maintain transparency in journalism. Media organizations must navigate the fine line between protecting sources and ensuring that the public remains informed about significant issues.

6. Case Studies: Quantum Communication in Action

Emerging initiatives are already beginning to demonstrate the potential of quantum communication in the media:

Secure Investigative Reporting: Some journalism organizations are piloting quantum communication systems for sensitive investigative reporting. By providing secure channels for communicating with whistleblowers, these initiatives enhance the integrity of the reporting process.

Quantum-Enabled News Aggregators: Innovative news aggregation platforms utilizing quantum algorithms are being developed to analyze and curate content from multiple sources, providing personalized news feeds while maintaining high standards of authenticity and accuracy.

Collaborative Journalism Projects: Collaborative journalism projects that rely on quantum communication to share information securely between journalists in different regions are emerging. This approach enhances the global reach of investigative reporting and fosters international cooperation.

The Future of Quantum Communication in Media

Quantum communication represents a significant leap forward in the evolution of journalism and media. By providing secure, rapid, and efficient communication channels, quantum technologies have the potential to enhance the integrity and reliability of news reporting while improving audience engagement.

As media organizations embrace quantum communication, it is crucial to prioritize ethical considerations, ensuring that the benefits of this technology are accessible to all and that public trust in journalism is upheld. By fostering collaboration between journalists, technologists, and policymakers, we can create a media landscape that leverages the power of quantum communication to inform, engage, and empower society.

CHAPTER 36:

The Quantum Singularity: The Limits of Human Knowledge

As quantum computing advances, it promises to solve problems that were once thought to be insurmountable, potentially leading humanity toward what can be termed a "quantum singularity." This chapter explores the philosophical implications of quantum computing's capabilities, addressing its potential to transcend human understanding, challenge the very foundations of knowledge, and provoke deep questions about existence, consciousness, and the nature of reality itself.

1. Defining the Quantum Singularity

The term "quantum singularity" evokes a scenario where quantum computers achieve a level of computational power that far exceeds traditional human reasoning and understanding. This could manifest in several ways:

Superintelligence: Quantum computers may be able to process and analyze vast datasets far beyond the capabilities of human cognition, leading to new forms of intelligence that operate on principles fundamentally different from our own.

Solving the Unsolvable: Many problems, such as the unification of general relativity and quantum mechanics, or the simulation of complex biological processes, remain unsolved. Quantum computers could potentially unlock solutions to these challenges, fundamentally altering our understanding of the universe.

Redefining Reality: As quantum computing reveals new insights into the fabric of reality, it could challenge long-held beliefs about existence, causality, and the nature of consciousness.

2. The Philosophical Landscape of Knowledge

To navigate the implications of the quantum singularity, we must first explore the nature of knowledge itself:

Epistemology and Quantum Computing: Epistemology, the study of knowledge, is fundamentally concerned with what we can know and how we can know it. Quantum computing challenges traditional epistemological frameworks, as it suggests that there may be realms of knowledge that are fundamentally inaccessible to human reasoning.

Limits of Human Cognition: Human cognition is inherently limited by our sensory perceptions, cognitive biases, and the constraints of classical reasoning. The emergence of quantum intelligence could amplify these limitations, as it operates on principles that may be beyond our comprehension.

Rethinking Truth and Reality: If quantum computing reveals truths about the universe that are counterintuitive or contradictory to human experience, we must grapple with the implications for our understanding of reality. Are there objective truths that exist independently of human perception, or is all knowledge inherently subjective?

3. Quantum Computing and the Nature of Problems

Quantum computers excel at solving specific types of problems that have stymied classical computers:

Complexity and Computation: Quantum computing's ability to leverage superposition and entanglement allows it to tackle problems with exponential complexity, such as factorization, optimization, and simulation. This capacity raises questions about the nature of computation itself and the limits of what problems can be solved.

Algorithmic Discovery: As quantum algorithms develop, they may discover solutions to problems that remain hidden from human reasoning. This potential presents profound questions about the role of human agency in knowledge production and problem-solving.

The Paradox of Understanding: If quantum computers can solve problems we cannot comprehend, what does this mean for our pursuit of knowledge? Do we redefine knowledge to include that which we cannot understand, or does true understanding require human cognition?

4. Implications for Ethics and Society

The emergence of quantum singularity raises significant ethical and societal considerations:

Ethical Decision-Making: If quantum systems can analyze vast datasets to recommend solutions for complex ethical dilemmas—such as climate change, healthcare, or social justice—should we trust these systems over human judgment? The implications for governance, policy-making, and social responsibility are profound.

Access and Inequality: The advent of quantum intelligence could create disparities between those who have access to quantum technologies and those who do not. This divide may exacerbate existing inequalities in knowledge, resources, and power.

The Role of Humans in Knowledge Creation: As quantum computing reshapes our understanding of knowledge, we must consider the role of human intuition, creativity, and emotion in knowledge production. Will machines ever replicate the nuances of human understanding, or will they exist as complementary entities?

5. Quantum Computing and Consciousness

The relationship between quantum computing and consciousness raises fascinating philosophical questions:

Quantum Consciousness: Some theories suggest that consciousness may be linked to quantum processes in the brain. If quantum computing can model or enhance aspects of consciousness, it could lead to breakthroughs in our understanding of the mind.

The Nature of Self: As we contemplate machines that think and reason beyond human capabilities, we must confront existential questions about the nature of self. What does it mean to be conscious, and how do we define individuality in a world where intelligent systems may exist independently of human experience?

Reality and Simulation: If quantum computers can simulate realities indistinguishable from our own, the line between reality and simulation becomes blurred. This raises questions about the nature of existence and the possibility of multiple, coexisting realities.

6. The Road Ahead: Embracing the Unknown

As we approach the quantum singularity, it is essential to embrace the unknown and remain open to new paradigms of thought:

Interdisciplinary Exploration: The challenges posed by the quantum singularity will require interdisciplinary collaboration across philosophy, science, technology, and ethics. By engaging in dialogue across fields, we can better navigate the complexities of knowledge and understanding.

Cultivating Quantum Literacy: As quantum technologies become more integrated into society, it is crucial to promote quantum literacy, empowering individuals to engage critically with these developments and their implications.

Navigating Uncertainty: Embracing uncertainty will be essential as we grapple with the profound implications of quantum computing. By acknowledging the limits of human knowledge and remaining curious about the unknown, we can foster a culture of exploration and inquiry.

Embracing a New Paradigm of Knowledge

The quantum singularity presents an unprecedented opportunity to redefine our understanding of knowledge, existence, and reality. As quantum computing continues to evolve, it challenges the very foundations of human cognition and invites us to explore the limits of what we can know.

By engaging thoughtfully with the philosophical implications of quantum computing, we can navigate this new landscape with purpose and intention. Embracing the potential of quantum technologies while acknowledging their challenges will be crucial in shaping a future where knowledge is expanded, understanding is deepened, and humanity continues to evolve in its quest for truth.

CHAPTER 37:

The Quantum Renaissance: A New Age of Discovery

As we stand on the threshold of unprecedented advancements in quantum technologies, we are poised to enter a new scientific renaissance—a transformative era characterized by groundbreaking discoveries that will reshape our understanding of the universe. This chapter explores how quantum technologies will revolutionize diverse fields, from physics to biology, igniting a spirit of innovation and inspiring future generations to explore the uncharted territories of knowledge.

1. Understanding the Quantum Renaissance

The term "quantum renaissance" encapsulates the transformative impact of quantum technologies on various scientific disciplines. This renaissance is marked by:

Interdisciplinary Collaboration: Quantum technologies will foster collaborations among physicists, chemists, biologists, engineers, and computer scientists, breaking down traditional barriers between fields and paving the way for holistic approaches to complex problems.

Innovation and Creativity: The quantum renaissance will encourage innovative thinking and creative problem-solving, inspiring researchers and entrepreneurs to explore new avenues of inquiry and develop novel solutions to pressing global challenges.

Cultural Shift: This period will cultivate a cultural shift that values curiosity, exploration, and interdisciplinary knowledge, encouraging individuals to embrace complexity and ambiguity in scientific pursuits.

2. Revolutionizing Physics: Unraveling the Universe's Mysteries

At the heart of the quantum renaissance lies the potential for revolutionary advancements in physics:

Quantum Field Theory and Beyond: Quantum technologies will enhance our understanding of fundamental particles and their interactions. New experiments, powered by quantum computing, may uncover deeper layers of reality, helping to unify general relativity with quantum mechanics and shedding light on phenomena like dark matter and dark energy.

Precision Measurement: Quantum sensors, leveraging the principles of superposition and entanglement, will enable unprecedented precision in measurements. This capability will enhance experimental physics, allowing for more accurate tests of theoretical predictions and advancing the search for new physics beyond the Standard Model.

Exploration of Quantum Gravity: The quantum renaissance will facilitate exploration into the elusive realm of quantum gravity, potentially leading to breakthroughs that challenge our understanding of spacetime itself and provide insights into the early moments of the universe.

3. Transforming Biology: The Quantum Approach to Life Sciences

Quantum technologies will extend their reach into biology, opening new frontiers in the life sciences:

Quantum Biology: Researchers are already exploring the quantum underpinnings of biological processes, such as photosynthesis and enzyme activity. Understanding these processes at the quantum level could lead to breakthroughs in renewable energy and sustainable technologies.

Drug Discovery and Personalized Medicine: Quantum computing will accelerate drug discovery by simulating molecular interactions and predicting the effectiveness of compounds. This capability promises to revolutionize personalized medicine, tailoring treatments to individual genetic profiles and improving patient outcomes.

Understanding Complex Biological Systems: Quantum simulations will enable researchers to model complex biological systems, including protein folding and cellular behavior. These insights could unlock new therapeutic targets and advance our understanding of diseases at the molecular level.

4. Inspiring Future Generations of Innovators

The quantum renaissance will not only transform existing fields but will also inspire a new generation of scientists, engineers, and innovators:

Education and Outreach: As quantum technologies become integrated into educational curricula, students will be exposed to cutting-edge concepts from an early age. Initiatives to promote quantum literacy will empower young minds to explore the possibilities of quantum science and technology.

Encouraging Creativity and Curiosity: By fostering a culture of curiosity and creativity, the quantum renaissance will inspire individuals to think outside the box, explore uncharted territories, and engage with the unknown. Encouraging interdisciplinary exploration will empower future innovators to tackle complex global challenges.

Creating New Career Opportunities: The growth of the quantum industry will generate new career paths in research, engineering, and technology development. As organizations seek talent with expertise in quantum computing, biology, and physics, a diverse range of professionals will be needed to drive innovation.

5. Addressing Global Challenges Through Quantum Technologies

The quantum renaissance has the potential to address some of the most pressing global challenges facing humanity:

Climate Change Mitigation: Quantum technologies will enhance our ability to model climate systems and optimize renewable energy solutions. By improving energy efficiency and enabling the development of new materials, quantum advancements could play a vital role in combating climate change.

Healthcare Innovation: The potential for quantum computing to revolutionize drug discovery, diagnostics, and treatment personalization offers transformative possibilities for healthcare systems. By improving patient outcomes and reducing costs, quantum technologies could reshape global health initiatives.

Sustainable Agriculture: Quantum technologies will enable more efficient agricultural practices, optimizing resource use and enhancing crop yields. These innovations can address food security challenges while promoting sustainable farming practices.

6. The Cultural Impact of the Quantum Renaissance

As the quantum renaissance unfolds, its cultural impact will be profound:

Art and Creativity: The interplay between quantum technologies and the arts will lead to new forms of artistic expression. Artists will explore quantum concepts, resulting in innovative works that bridge science and creativity, inspiring wonder and curiosity.

Public Engagement with Science: The excitement surrounding quantum technologies will engage the public in scientific discourse, fostering a deeper understanding of complex concepts and promoting informed discussions about the implications of these advancements.

Ethical Considerations: As new technologies emerge, ethical considerations will arise regarding their applications and implications for society. Engaging in discussions about the responsible use of quantum technologies will be essential to navigate potential challenges and ensure equitable access.

Embracing the Quantum Renaissance

The quantum renaissance heralds a new age of discovery, offering unprecedented opportunities for innovation across various scientific disciplines. By revolutionizing fields such as physics and biology, quantum technologies will reshape our understanding of the universe and inspire future generations to explore the unknown.

As we embrace this new era, it is crucial to foster a culture of curiosity, collaboration, and creativity, empowering individuals to engage with quantum concepts and contribute to the unfolding narrative of discovery. The potential of quantum technologies is vast, and their impact on humanity will be profound.

CHAPTER 38:

Preparing for the Quantum Future

As we move closer to a world transformed by quantum computing, it becomes essential for individuals, organizations, and societies to proactively prepare for the changes and opportunities this technology will bring. This chapter outlines practical strategies for navigating the quantum future, focusing on corporate strategies, educational shifts, and personal growth.

1. Understanding the Quantum Landscape

To effectively prepare for the quantum future, it is essential to develop a foundational understanding of quantum computing and its implications:

Stay Informed: Keeping abreast of advancements in quantum technologies, research breakthroughs, and industry trends will empower individuals and organizations to make informed decisions. Subscribing to reputable journals, attending conferences, and following thought leaders in the field will enhance awareness.

Explore Quantum Concepts: Familiarizing oneself with key quantum principles—such as superposition, entanglement, and quantum algorithms—will help demystify the technology. Online courses, webinars, and educational resources can provide accessible entry points for understanding quantum computing.

Engage with Quantum Communities: Joining online forums, participating in meetups, and engaging with academic institutions or tech organizations focused on quantum computing will foster connections and knowledge exchange.

2. Corporate Strategies for Quantum Adoption

Organizations must develop robust strategies to leverage quantum computing effectively:

Assess Business Needs: Identify areas within the organization where quantum computing could provide a competitive advantage. This might include optimization, data analysis, cryptography, or simulation tasks that require immense computational power.

Invest in Quantum Talent: Building a workforce skilled in quantum computing is crucial for future success. Companies should prioritize hiring talent with expertise in quantum algorithms, physics, and computer science, and invest in training programs to upskill existing employees.

Collaborate with Quantum Startups: Partnering with quantum startups and research institutions can accelerate innovation. Collaboration can provide access to cutting-edge technologies, tools, and expertise that enhance organizational capabilities.

Pilot Quantum Projects: Organizations should initiate pilot projects to experiment with quantum computing applications. By starting small and scaling successful initiatives, companies can gradually integrate quantum solutions into their operations.

3. Educational Shifts for the Quantum Era

The educational system must evolve to equip future generations with the skills and knowledge necessary for thriving in a quantum-driven world:

Revise Curricula: Educational institutions should integrate quantum computing concepts into STEM curricula at all levels. This includes introducing quantum principles in primary education and offering specialized programs in higher education focused on quantum technologies.

Promote Interdisciplinary Learning: Encouraging interdisciplinary approaches that blend quantum science with fields such as ethics, social sciences, and the arts will foster a holistic understanding of the implications of quantum technologies.

Enhance Accessibility: Making quantum education accessible to diverse populations is essential for equitable participation in the quantum future. Online learning platforms, community workshops, and outreach programs can democratize knowledge and encourage broader engagement.

Encourage Critical Thinking: As quantum computing raises complex ethical and societal questions, cultivating critical thinking skills among students will be vital. Teaching students to navigate ambiguity and consider the implications of technology will prepare them for future challenges.

4. Personal Growth in a Quantum World

Individuals can take proactive steps to grow and adapt in a world increasingly influenced by quantum technologies:

Cultivate a Growth Mindset: Embracing a growth mindset—believing in the ability to learn and adapt—will be crucial for personal development. Openness to new ideas and experiences will enable individuals to thrive in the dynamic landscape of quantum computing.

Invest in Lifelong Learning: Committing to lifelong learning will be essential for staying relevant in the workforce. Engaging in continuous education, attending workshops, and pursuing certifications in quantum computing or related fields will enhance career prospects.

Develop Soft Skills: In addition to technical skills, cultivating soft skills such as communication, collaboration, and problem-solving will be crucial for success in a quantum future. These skills will enable individuals to work effectively in interdisciplinary teams and navigate complex challenges.

Engage with Ethical Discussions: As quantum technologies raise ethical considerations, individuals should engage in discussions about the implications of their applications. Understanding the societal impact of quantum computing will empower individuals to advocate for responsible use and contribute to shaping ethical frameworks.

5. Preparing for Quantum Innovations in Daily Life

As quantum technologies become integrated into everyday life, individuals will need to adapt their daily practices:

Embrace Quantum Solutions: Be open to adopting quantum-powered applications and services that enhance daily life, from improved healthcare diagnostics to optimized transportation systems. Understanding the benefits of quantum innovations will enable informed choices.

Prioritize Cybersecurity Awareness: With the rise of quantum computing, awareness of cybersecurity measures will be vital. Individuals should educate themselves about quantum-safe encryption methods and protect personal data against emerging threats.

Promote Sustainability: Quantum technologies offer opportunities for sustainability in various sectors, including energy, agriculture, and manufacturing. Individuals can advocate for and support initiatives that leverage quantum innovations to promote environmental stewardship.

Embracing the Quantum Future

Preparing for the quantum future requires a proactive, multifaceted approach that encompasses individual growth, corporate strategies, and educational shifts. By staying informed about quantum technologies, organizations can position themselves to leverage the advantages they offer, while individuals can cultivate the skills and knowledge necessary to thrive in this evolving landscape.

As we approach this new era, the possibilities are vast, and the implications are profound. Embracing the quantum future means embracing curiosity, collaboration, and a commitment to lifelong learning. It challenges us to think critically about the intersection of technology, ethics, and society.

CHAPTER 39:

The Dark Side of Quantum Computing

As we delve deeper into the transformative realm of quantum computing, it is essential to confront the potential dangers that accompany this powerful technology. While quantum computing holds immense promise for solving complex problems and driving innovation, it also poses significant risks that could lead to economic disruption, ethical misuse, and the empowerment of malicious actors. This chapter explores the dark side of quantum computing, highlighting the challenges and dangers that society must address to ensure a safe and responsible transition into the quantum era.

1. Economic Disruption: The Quantum Divide

The advent of quantum computing has the potential to exacerbate existing inequalities and create a new economic divide:

Job Displacement: As organizations adopt quantum computing technologies to automate processes and optimize operations, there is a risk of widespread job displacement. Traditional roles in data analysis, cybersecurity, and cryptography may become obsolete as quantum algorithms outperform classical methods.

Market Instability: The rapid emergence of quantum computing could destabilize financial markets. Companies that fail to adopt quantum technologies may struggle to compete, leading to market volatility and potential bankruptcies. Conversely, those that successfully leverage quantum computing may gain disproportionate advantages, creating winners and losers in the economic landscape.

Access to Quantum Resources: The concentration of quantum computing capabilities in wealthier nations and corporations may further entrench global inequalities. Countries or organizations that lack access to quantum resources could fall behind in technological advancements, limiting their economic growth and development.

2. Ethical Misuse: A Double-Edged Sword

Quantum computing's capabilities can be a double-edged sword, opening avenues for both innovation and misuse:

Cryptographic Vulnerabilities: One of the most pressing concerns surrounding quantum computing is its potential to break widely used encryption methods. Quantum algorithms, such as Shor's algorithm, could render current cryptographic systems obsolete, exposing sensitive information and leading to widespread data breaches.

Surveillance and Privacy Erosion: Quantum computing could enable advanced surveillance techniques, allowing malicious actors or authoritarian regimes to monitor citizens more effectively. This erosion of privacy could have chilling effects on freedom of expression and individual rights.

Biased Algorithms: As quantum computing intersects with artificial intelligence, there is a risk that biased data or algorithms could lead to discriminatory outcomes. If not carefully managed, quantum-enhanced AI could perpetuate existing societal inequalities and biases, exacerbating issues of fairness and justice.

3. Empowering Malicious Actors: The Weaponization of Quantum Technologies

The rise of quantum computing presents significant risks in terms of malicious use, particularly in the context of cybersecurity and warfare:

Cybersecurity Threats: Malicious actors could harness quantum computing to launch sophisticated cyberattacks. Quantum-enabled hacking tools could breach secure systems, steal sensitive information, and disrupt critical infrastructure. The potential for quantum-powered cyber warfare poses a grave threat to national security and global stability.

Weaponization of Quantum Technologies: As nations race to achieve quantum supremacy, there is a risk that quantum technologies could be weaponized for offensive purposes. The development of quantum-enhanced weapons or surveillance systems could escalate geopolitical tensions and lead to an arms race in quantum technologies.

Terrorism and Organized Crime: Criminal organizations may leverage quantum computing to enhance their capabilities, from planning sophisticated heists to executing cyberattacks. The accessibility of quantum technologies could empower malicious actors and increase the sophistication of criminal enterprises.

4. The Regulatory Challenge: Ensuring Responsible Use

Addressing the dark side of quantum computing requires robust regulatory frameworks and proactive governance:

Establishing Ethical Guidelines: Governments and organizations must work together to establish ethical guidelines for the development and use of quantum technologies. These guidelines should address issues such as data privacy, algorithmic bias, and the responsible use of quantum computing in sensitive areas.

International Collaboration: Given the global nature of quantum computing, international cooperation will be essential in regulating its use. Collaborative efforts among nations can help establish norms and agreements to prevent the weaponization of quantum technologies and mitigate cybersecurity threats.

Promoting Transparency: Transparency in quantum research and applications is crucial for fostering trust and accountability. Open dialogue about the potential risks and ethical implications of quantum computing will empower stakeholders to address concerns collaboratively.

5. Preparing for the Dark Side: Mitigation Strategies

To mitigate the potential dangers of quantum computing, individuals, organizations, and governments can take proactive measures:

Invest in Quantum-Safe Cryptography: As the threat of quantum computing breaking current encryption methods looms, investing in quantum-safe cryptographic techniques is essential. Developing and adopting new standards for secure communication will be critical in safeguarding sensitive data.

Enhance Cybersecurity Measures: Organizations must bolster their cybersecurity frameworks to withstand potential quantum-enabled attacks. This includes adopting multi-layered security approaches, conducting regular assessments, and staying informed about emerging threats.

Foster Public Awareness and Education: Raising public awareness about the implications of quantum computing will empower individuals to engage in informed discussions about its risks and benefits. Educational initiatives can help demystify quantum technologies and promote critical thinking about their ethical implications.

Navigating the Quantum Landscape Responsibly

As we venture into the quantum future, it is crucial to acknowledge the potential dangers that accompany this transformative technology. While quantum computing offers remarkable opportunities for innovation and progress, we must confront the dark side and work collaboratively to mitigate its risks.

By establishing ethical guidelines, promoting transparency, and investing in cybersecurity measures, we can navigate the quantum landscape responsibly and ensure that the benefits of quantum computing are realized without compromising safety, privacy, and equality.

CHAPTER 40:

Quantum Ecosystems: Building the Infrastructure

As quantum computing moves from the realm of theoretical research into practical application, the establishment of a robust quantum ecosystem becomes essential. This ecosystem will not only facilitate the development and deployment of quantum technologies but will also ensure that they are accessible, sustainable, and beneficial to society as a whole. This chapter explores the critical components of a quantum ecosystem, including hardware and software infrastructure, quantum cloud services, and vibrant developer communities.

1. The Need for a Quantum Ecosystem

The complexities and unique requirements of quantum computing necessitate a specialized infrastructure that supports both developers and end-users:

Specialized Hardware: Quantum computers rely on advanced hardware, including qubits, superconducting circuits, and cryogenic systems, which are fundamentally different from classical computing components. Developing and manufacturing these components requires significant investment and expertise.

Software Frameworks: Quantum algorithms and applications require dedicated software frameworks that can efficiently interface with quantum hardware. Developing programming languages and development environments tailored to quantum computing is essential for enabling developers to write and optimize quantum algorithms.

Supportive Infrastructure: Beyond just hardware and software, a comprehensive infrastructure is needed to support the deployment, testing, and scaling of quantum applications. This includes integration with existing classical systems and the establishment of standards for interoperability.

2. Quantum Hardware: The Backbone of the Ecosystem

The hardware landscape for quantum computing is diverse and continuously evolving. Key aspects include:

Qubit Technologies: Various technologies, such as superconducting qubits, trapped ions, and topological qubits, are being researched and developed. Each has its advantages and challenges, and the choice of technology will significantly influence the performance and scalability of quantum computers.

Manufacturing and Scaling: Building quantum hardware at scale presents unique challenges. Developing manufacturing processes that can produce reliable and consistent qubits while maintaining performance is critical. As demand for quantum solutions grows, innovations in manufacturing techniques will be essential.

Maintenance and Stability: Quantum computers require precise conditions to function optimally. Ensuring the stability and reliability of quantum hardware, including maintaining cryogenic temperatures and minimizing decoherence, is crucial for long-term success.

3. Software Development: Bridging the Quantum-Classical Divide

As quantum computing becomes more practical, a robust software ecosystem is needed to facilitate the development of quantum applications:

Quantum Programming Languages: Developing specialized programming languages, such as Qiskit, Cirq, and Quipper, will enable developers to express quantum algorithms efficiently. These languages must be intuitive and provide access to quantum gates and operations that represent quantum phenomena accurately.

Compilers and Optimization Tools: Quantum compilers that translate high-level quantum algorithms into low-level instructions compatible with quantum hardware are essential. Optimization tools that can refine these algorithms for performance and resource efficiency will also play a crucial role.

Integration with Classical Systems: Creating hybrid systems that leverage both classical and quantum computing will be vital. Many real-world applications will require seamless integration between traditional computing resources and quantum capabilities, necessitating the development of compatible APIs and communication protocols.

4. Quantum Cloud Services: Democratizing Access

Quantum cloud computing services are vital for expanding access to quantum technologies, especially for smaller organizations and individuals:

Cloud-Based Quantum Platforms: Quantum cloud services allow users to access quantum computers remotely, enabling experimentation without the need for significant upfront investment in hardware. Platforms such as IBM Quantum Experience and Amazon Braket provide users with tools to run quantum algorithms and simulate quantum circuits.

Cost-Effective Solutions: By offering pay-per-use models, quantum cloud services reduce the barrier to entry for organizations and researchers, allowing them to explore quantum applications without committing to expensive hardware purchases.

Community Engagement: Cloud platforms often foster user communities that share knowledge, tools, and best practices. These communities can facilitate collaboration and accelerate the development of innovative quantum applications.

5. Developer Communities: Fostering Collaboration and Innovation

A vibrant developer community is crucial for the growth and sustainability of the quantum ecosystem:

Educational Initiatives: Online courses, workshops, and hackathons can engage developers and researchers, enhancing their understanding of quantum computing and encouraging experimentation with quantum algorithms.

Open Source Contributions: Promoting open-source development will enable collaboration and innovation within the quantum computing community. Open-source libraries, frameworks, and tools can accelerate progress and reduce duplication of effort.

Industry Partnerships: Collaborations between academia, industry, and government can foster a collaborative ecosystem that supports knowledge transfer and innovation. These partnerships can lead to advancements in quantum research, application development, and workforce training.

6. Regulatory and Ethical Considerations: Building a Responsible Ecosystem

As the quantum ecosystem expands, it is crucial to address regulatory and ethical considerations:

Standards and Protocols: Establishing industry standards for quantum hardware, software, and security will ensure interoperability and reliability. Collaborative efforts among stakeholders can help develop these standards, fostering a cohesive ecosystem.

Ethical Guidelines: The development of ethical guidelines for quantum research and applications will be essential to ensure responsible use. Stakeholders must consider the implications of quantum technologies on privacy, security, and societal equity.

Policy Support: Governments and regulatory bodies should actively support the development of quantum ecosystems through funding, research grants, and policy initiatives that encourage innovation while safeguarding public interests.

Cultivating a Thriving Quantum Ecosystem

Building a robust quantum ecosystem is vital for unlocking the full potential of quantum computing. By investing in specialized hardware, developing tailored software frameworks, expanding access through quantum cloud services, and fostering collaborative developer communities, we can create an environment conducive to innovation and progress.

CHAPTER 41:

The Role of Tech Giants and Startups

The landscape of quantum computing is being profoundly shaped by the combined efforts of established tech giants, agile startups, and prominent research institutions. Each of these entities plays a crucial role in advancing the technology, driving innovation, and democratizing access to quantum computing resources. This chapter explores the significant contributions of these players, their competitive dynamics, and their collaborative efforts to propel quantum computing into a mainstream reality.

1. Tech Giants: The Powerhouses of Quantum Innovation

Major technology companies have recognized the immense potential of quantum computing and are investing heavily in its development. Their substantial resources enable them to spearhead research, build advanced hardware, and create robust software ecosystems.

IBM: As one of the pioneers in quantum computing, IBM has been at the forefront of developing quantum hardware and software. The IBM Quantum Experience platform allows researchers and developers to access their quantum computers via the cloud, fostering a community of quantum enthusiasts. IBM's commitment to quantum research is underscored by its ambitious goal of achieving "quantum advantage" through the development of practical quantum applications.

Google: Google made headlines in 2019 when its Sycamore processor reportedly achieved quantum supremacy by performing a specific calculation much faster than the best classical supercomputers. Google continues to advance quantum computing research through its Quantum AI lab, focusing on developing new quantum algorithms and hardware innovations that could solve real-world problems.

Microsoft: Microsoft's approach to quantum computing includes its Quantum Development Kit, which enables developers to write quantum programs using Q#. The company is investing in building a scalable quantum computing platform based on topological qubits, which promise greater stability and error correction.

Amazon: Amazon's entry into quantum computing is facilitated through Amazon Braket, a cloud-based service that allows users to explore and experiment with quantum algorithms on different types of quantum hardware. By providing scalable and accessible quantum solutions, Amazon is democratizing access to quantum technology for researchers and businesses alike.

2. Startups: The Agile Innovators

While tech giants have the advantage of resources and infrastructure, startups are often more nimble, allowing them to experiment with innovative ideas and approaches to quantum computing:

Rigetti Computing: Rigetti focuses on building quantum computers and a software ecosystem to enable quantum applications. Its cloud-based Quantum Cloud Services (QCS) allows developers to access its quantum processors and integrate them into their workflows, promoting broader adoption of quantum technologies.

IonQ: Specializing in trapped-ion quantum computing, IonQ is committed to developing scalable quantum systems. The company emphasizes the importance of building quantum computers that can be readily accessed via the cloud, aiming to make quantum computing accessible to a wider audience.

D-Wave Systems: D-Wave is a leader in quantum annealing technology, which focuses on optimization problems. The company has made strides in creating hybrid systems that combine classical and quantum computing to address real-world challenges, including supply chain optimization and machine learning.

Quantum Machines: This startup is focused on building the orchestration software required for quantum computers to operate effectively. Quantum Machines aims to provide a complete stack that simplifies the integration and management of quantum processors, enhancing usability for developers and researchers.

3. Research Institutions: The Foundation of Quantum Advancement

In addition to corporate players, research institutions and universities are critical to advancing quantum computing through fundamental research, experimentation, and education:

MIT and Harvard: These institutions are at the forefront of quantum research, exploring various aspects of quantum theory, quantum information science, and practical applications. Their collaborations often lead to groundbreaking discoveries and the development of new quantum algorithms and materials.

Caltech: Caltech's Institute for Quantum Information and Matter (IQIM) is dedicated to understanding and harnessing quantum phenomena for practical applications. The institute fosters interdisciplinary collaborations among physicists, engineers, and computer scientists to drive innovation in quantum technologies.

University of Waterloo: The Institute for Quantum Computing (IQC) at the University of Waterloo is a global leader in quantum research. Its collaborative approach brings together experts from various fields to tackle the challenges associated with quantum computing and its applications.

4. Collaborations and Partnerships: Strength in Unity

The complexity of quantum computing necessitates collaboration among tech giants, startups, and research institutions. These partnerships can accelerate progress and foster innovation:

Joint Research Initiatives: Collaborative research projects can leverage the strengths of different organizations. For example, partnerships between universities and tech companies can facilitate access to cutting-edge quantum hardware while providing a rich pool of academic talent.

Industry-Academia Collaborations: Programs that bridge the gap between academia and industry can enhance educational outcomes and create pathways for students to engage with quantum computing technologies. Internships, co-op programs, and joint research initiatives can cultivate a new generation of quantum talent.

Shared Resources: Collaborative initiatives may involve sharing quantum resources, such as cloud access to quantum computers or joint development of software tools. This collective approach can foster innovation and lower costs, making quantum computing more accessible to a broader audience.

5. The Competitive Landscape: Innovation vs. Collaboration

The competitive dynamics between tech giants and startups can drive innovation in quantum computing. However, this competition must be balanced with collaboration to ensure the technology's responsible development:

Intellectual Property Concerns: As companies race to develop proprietary quantum technologies, concerns regarding intellectual property rights can arise. Striking a balance between protecting innovations and fostering collaboration will be essential for long-term success.

Market Fragmentation: The proliferation of quantum solutions can lead to market fragmentation, where various systems and protocols hinder interoperability. Establishing common standards and frameworks will be crucial to ensuring that quantum technologies can work together seamlessly.

Ethical Responsibility: As major players in the quantum landscape, tech giants and startups alike have a responsibility to consider the ethical implications of their work. Transparent communication about research goals and potential risks will be vital in building public trust and ensuring that quantum technologies are developed responsibly.

Shaping the Future of Quantum Computing Together

The future of quantum computing is being shaped by a diverse array of players, each contributing unique strengths and perspectives. Tech giants provide the resources and infrastructure necessary for large-scale development, while startups introduce agility and innovation to the landscape. Research institutions lay the groundwork for scientific discovery and education.

CHAPTER 42:

Quantum Computing and Human Augmentation

As quantum computing continues to evolve, its potential to revolutionize biotechnology and human augmentation is becoming increasingly apparent. This chapter explores how quantum technologies could enhance human cognitive and physical abilities, fundamentally transforming our understanding of what it means to be human. By integrating quantum computing with advancements in biotechnology, we can pave the way for unprecedented innovations in health, enhancement, and human experience.

1. Understanding Human Augmentation

Human augmentation refers to the use of technology to enhance human physical and cognitive capabilities. This encompasses a wide range of applications, including:

Cognitive Enhancement: Techniques aimed at improving memory, learning, and decision-making processes. This could involve neurotechnologies, such as brain-computer interfaces (BCIs), that enhance mental capabilities.

Physical Enhancement: Technologies that improve physical abilities, such as exoskeletons that enhance strength or mobility, or biomedical implants that restore or enhance bodily functions.

Health Monitoring and Improvement: Continuous monitoring of health metrics through wearable devices and personalized medicine tailored to individual genetic profiles.

2. The Role of Quantum Computing in Biotechnology

Quantum computing has the potential to significantly advance biotechnology through its ability to process vast amounts of data and solve complex problems that are currently beyond the reach of classical computers:

Drug Discovery and Development: Quantum algorithms can simulate molecular interactions at an unprecedented scale, enabling researchers to identify potential drug candidates more rapidly and accurately. This accelerates the drug discovery process, making it possible to develop targeted therapies tailored to individual patients.

Genomic Analysis: The human genome is incredibly complex, and analyzing genetic data requires immense computational power. Quantum computing can facilitate faster and more accurate genomic sequencing, enabling personalized medicine that tailors treatments based on an individual's genetic makeup.

Protein Folding: Understanding how proteins fold into their functional shapes is crucial for many biological processes. Quantum simulations can model protein folding dynamics more effectively than classical methods, potentially leading to breakthroughs in disease understanding and treatment.

3. Enhancing Cognitive Abilities

Quantum computing's power extends to enhancing human cognition by improving our understanding of the brain and developing technologies that augment mental capabilities:

Brain-Computer Interfaces (BCIs): BCIs enable direct communication between the brain and external devices, allowing for enhanced control of technology and even the potential for cognitive enhancements. Quantum computing can enhance the data processing capabilities of BCIs, improving their responsiveness and accuracy.

Neural Networks and Machine Learning: Quantum computing can accelerate machine learning algorithms used in neuroscience research, facilitating the analysis of brain activity data. This could lead to better understanding and treatment of cognitive disorders and the development of personalized cognitive enhancement strategies.

Enhanced Learning Systems: With quantum algorithms that optimize learning processes, educational systems could be transformed. Personalized learning experiences based on real-time data analytics can tailor educational content to individual needs, potentially boosting cognitive performance.

4. Physical Augmentation: Merging Quantum and Biomechanics

Quantum computing's capabilities could also revolutionize physical augmentation, leading to the development of sophisticated devices and systems that enhance human abilities:

Exoskeletons and Prosthetics: Quantum computing could optimize the design and functionality of exoskeletons and prosthetics, allowing for more intuitive control and better integration with the human body. Quantum algorithms could simulate complex movements, ensuring that these devices respond seamlessly to user intentions.

Bioengineering and Tissue Regeneration: The ability to model biological systems at a quantum level may facilitate breakthroughs in tissue engineering and regenerative medicine. Quantum technologies could assist in creating custom-designed tissues and organs that integrate perfectly with the human body, enhancing healing processes and restoring lost functions.

Wearable Technology: Quantum computing could lead to the development of advanced wearable technologies that monitor health metrics in real time and adapt to changing conditions. These wearables could provide immediate feedback, enhancing athletic performance and promoting overall well-being.

5. Ethical Considerations in Human Augmentation

As we explore the intersection of quantum computing and human augmentation, ethical considerations become paramount:

Equity and Access: Ensuring that advancements in human augmentation are accessible to all individuals, regardless of socioeconomic status, is critical. The risk of creating a divide between those who can afford enhancements and those who cannot must be addressed to avoid societal inequalities.

Consent and Autonomy: The use of technologies that enhance human capabilities raises questions about consent and personal autonomy. Individuals must have the right to choose whether to augment themselves and understand the implications of those choices.

Identity and Humanity: The integration of technology into human biology challenges traditional notions of identity and what it means to be human. As augmentation technologies advance, society must grapple with the philosophical implications of enhancing human abilities and the potential impact on human experience.

6. The Future of Human Augmentation in a Quantum World

The future of human augmentation, powered by quantum computing, promises remarkable possibilities:

New Frontiers in Health: Enhanced capabilities in drug discovery, genomics, and personalized medicine could lead to breakthroughs in treating diseases that are currently untreatable, improving health outcomes for millions.

Empowered Individuals: Augmented cognitive and physical abilities could empower individuals to achieve more, whether in professional settings or personal pursuits. This could lead to new forms of creativity, problem-solving, and collaboration.

Rethinking Society: As human augmentation becomes more prevalent, society may need to redefine its values, norms, and systems to accommodate these changes. Discussions about the ethical implications of augmentation will be crucial in shaping a future that respects individual rights and collective well-being.

CHAPTER 43:

Quantum Religion: Spirituality in a Quantum World

As quantum computing and quantum mechanics continue to unfold, they not only reshape our understanding of the physical universe but also provoke profound questions about spirituality, consciousness, and the nature of existence. This chapter explores how the principles of quantum theory might influence spiritual beliefs and philosophical outlooks, offering new ways to comprehend humanity's relationship with the unknown.

1. Quantum Mechanics: A New Lens for Spirituality

Quantum mechanics challenges classical perceptions of reality, introducing concepts that can resonate deeply with spiritual philosophies. These include:

Interconnectedness: One of the foundational principles of quantum mechanics is entanglement, where particles become interconnected regardless of distance. This idea mirrors many spiritual traditions that emphasize unity and the interconnectedness of all beings, suggesting that everything in the universe is intertwined in ways we have yet to fully understand.

Observer Effect: The observer effect in quantum mechanics posits that the act of observation can alter the behavior of a particle. This notion raises intriguing questions about consciousness and the role of the observer, paralleling spiritual teachings that highlight the importance of awareness and intention in shaping our experiences.

Non-Locality: Quantum entanglement challenges the classical notion of locality, indicating that information can be shared instantaneously across distances. This notion could parallel spiritual beliefs in telepathy or collective consciousness, suggesting that human minds might be more interconnected than we perceive.

2. Philosophical Implications of Quantum Discoveries

The revelations of quantum physics invite a reevaluation of philosophical outlooks concerning existence, reality, and the nature of the universe:

Reality as a Construct: Quantum mechanics suggests that reality may not be as fixed and objective as we once thought. This perspective aligns with various philosophical and spiritual traditions that view reality as a construct influenced by consciousness, encouraging individuals to explore the nature of their perceptions and beliefs.

The Nature of Consciousness: Quantum theories have prompted discussions about consciousness, suggesting that it may play a fundamental role in shaping reality. This opens the door to exploring the relationship between consciousness and spirituality, leading to questions about the soul, awareness, and the afterlife.

Meaning and Purpose: As we confront the implications of quantum discoveries, individuals may begin to seek new meanings and purposes. Spiritual philosophies often emphasize finding meaning in existence, and quantum mechanics may provide a scientific framework to support these inquiries.

3. Bridging Science and Spirituality

The intersection of quantum mechanics and spirituality provides an opportunity for dialogue between science and spiritual beliefs:

Complementary Perspectives: Instead of viewing science and spirituality as opposing forces, they can be seen as complementary approaches to understanding existence. Quantum discoveries may enrich spiritual practices, offering insights into the interconnectedness of life and the universe.

Meditation and Consciousness: Practices such as meditation, which aim to enhance awareness and connect with deeper aspects of consciousness, can be explored through a quantum lens. Understanding the observer's role may encourage a deeper appreciation for practices that cultivate mindfulness and presence.

New Paradigms: As quantum mechanics continues to challenge traditional scientific paradigms, spiritual thinkers may draw inspiration from these shifts, proposing new frameworks that encompass both scientific inquiry and spiritual exploration.

4. The Search for Universal Truths

Quantum mechanics invites individuals to engage in a search for universal truths that transcend specific doctrines and belief systems:

Global Spiritual Movements: As people explore the implications of quantum theory, there may be a rise in spiritual movements that emphasize unity, interconnectedness, and the exploration of consciousness. These movements could draw from various traditions, fostering a global dialogue on spirituality.

Holistic Understanding: Emphasizing a holistic approach to spirituality can encourage individuals to explore various paths while recognizing the underlying unity that connects all beliefs. This approach aligns with the notion that different spiritual traditions may ultimately seek the same truths.

5. Challenges and Controversies

While the relationship between quantum mechanics and spirituality offers exciting possibilities, it also raises challenges and controversies:

Misinterpretation of Quantum Concepts: The popularity of quantum mechanics in spiritual discourse can sometimes lead to misinterpretations or sensationalism. It's essential to approach these discussions with a nuanced understanding of both quantum theory and spiritual beliefs to avoid oversimplifications.

Scientific Rigor vs. Spiritual Exploration: The interplay between rigorous scientific inquiry and subjective spiritual experiences can create tensions. Striking a balance between empirical evidence and personal belief is crucial in navigating this complex territory.

6. The Future of Quantum Spirituality

As quantum computing and quantum mechanics continue to develop, their influence on spirituality and belief systems is likely to expand:

Emerging Discourses: New spiritual discourses may arise that explore the implications of quantum discoveries, integrating scientific insights into traditional spiritual practices and philosophies.

Interdisciplinary Collaborations: Future explorations may involve collaborations between scientists, philosophers, and spiritual leaders, creating interdisciplinary dialogues that enrich our understanding of consciousness and existence.

Personal Journeys: As individuals engage with quantum ideas, they may embark on personal journeys of exploration, seeking to reconcile their scientific understanding with their spiritual beliefs. This journey could lead to greater self-awareness and a deeper connection to the universe.

CHAPTER 44:

Quantum Governance: The Global Rulebook

As quantum technologies rapidly advance, the need for effective governance structures becomes increasingly urgent. These technologies, while promising significant benefits across multiple sectors, also raise complex ethical, security, and social equity concerns. This chapter proposes new governance models designed to ensure the responsible development and use of quantum technologies, addressing issues such as global security, privacy, and equitable access.

1. The Need for Quantum Governance

The unique characteristics of quantum technologies necessitate a fresh approach to governance. Key factors include:

Rapid Technological Advancement: Quantum computing and related technologies are evolving at an unprecedented pace, often outstripping existing regulatory frameworks. This speed creates challenges for policymakers seeking to understand and manage their implications.

Global Security Concerns: Quantum computing has the potential to disrupt existing security paradigms, particularly in fields like cybersecurity and international relations. Quantum capabilities could render current encryption methods obsolete, raising concerns about data integrity and national security.

Ethical and Social Implications: The advent of quantum technologies introduces new ethical dilemmas, including issues related to privacy, inequality in access, and the potential misuse of powerful quantum capabilities.

2. Principles of Quantum Governance

Effective governance models must be founded on principles that promote responsible and equitable development:

Transparency: Decision-making processes regarding quantum technologies should be transparent to build public trust. Stakeholders, including the general public, must be informed about the implications of quantum advancements.

Inclusivity: Governance frameworks should include diverse voices, particularly those from marginalized communities, ensuring that the benefits of quantum technologies are distributed equitably.

Adaptability: Given the rapid evolution of quantum technologies, governance models must be flexible and capable of adapting to new developments and unforeseen challenges.

Global Collaboration: As quantum technologies transcend national boundaries, international cooperation is crucial. Collaborative governance frameworks can help address global security concerns and establish common standards.

3. Proposed Governance Models

To address the unique challenges posed by quantum technologies, several governance models can be considered:

International Regulatory Bodies: Establishing international organizations dedicated to quantum governance can facilitate cooperation among nations. These bodies could develop global standards for quantum research, security protocols, and ethical guidelines, similar to frameworks established for nuclear safety.

Public-Private Partnerships: Collaborations between governments, academic institutions, and private companies can ensure that quantum technologies are developed responsibly. Public-private partnerships can foster innovation while maintaining accountability and oversight.

Quantum Ethics Committees: Creating specialized ethics committees to evaluate quantum research proposals and applications can help address ethical concerns before they arise. These committees could provide guidance on issues such as privacy, consent, and the societal impact of quantum technologies.

Dynamic Regulatory Frameworks: Rather than static regulations, dynamic frameworks that evolve with technological advancements can better accommodate the fast-paced nature of quantum innovation. These frameworks could include mechanisms for continuous assessment and adjustment.

4. Addressing Global Security Concerns

Quantum technologies have the potential to reshape global security landscapes, particularly in cybersecurity:

Quantum Cybersecurity Strategies: Developing quantum-resistant encryption methods is essential to protect data in the quantum era. Governance frameworks should prioritize research and funding for quantum cybersecurity initiatives to safeguard sensitive information.

International Treaties: Just as arms control treaties seek to prevent the proliferation of nuclear weapons, international agreements could be established to regulate the development and use of quantum technologies for military purposes. Such treaties can help mitigate the risk of a quantum arms race.

Security Protocols for Quantum Communication: As quantum communication systems emerge, establishing security protocols to protect data transmitted through these networks will be crucial. Governance frameworks must prioritize the development of these protocols to ensure data integrity and confidentiality.

5. Ensuring Privacy and Social Equity

Quantum governance must also address privacy concerns and the risk of exacerbating social inequalities:

Privacy Regulations: As quantum technologies enable unprecedented data analysis capabilities, robust privacy regulations are needed to protect individuals' rights. Laws should ensure that quantum advancements do not compromise personal privacy or consent.

Equitable Access Initiatives: To prevent a quantum divide, governance models should promote equitable access to quantum technologies. Funding and support for educational programs, research grants, and infrastructure development in underserved communities can help bridge the gap.

Social Impact Assessments: Before deploying quantum technologies, conducting social impact assessments can identify potential risks and benefits. These assessments should consider how technologies may affect various communities and ensure that marginalized groups are not disproportionately impacted.

6. Fostering Global Collaboration

Given the transnational nature of quantum technologies, fostering global collaboration is essential:

Multilateral Dialogues: Hosting international conferences and forums dedicated to quantum governance can facilitate discussions among governments, researchers, and industry leaders. These dialogues can help establish shared goals and best practices.

Knowledge Sharing Platforms: Creating platforms for knowledge sharing can encourage collaboration among researchers and institutions across borders. Open access to research findings and data can accelerate innovation while promoting responsible development.

Collaborative Research Initiatives: Joint research projects involving multiple nations can address common challenges in quantum technologies. These initiatives can facilitate knowledge exchange and promote collective problem-solving.

CHAPTER 45:

Quantum Leadership: Innovators of Tomorrow

As we stand on the brink of a quantum revolution, a new cadre of leaders is emerging, poised to shape the future of technology, science, and society. These innovators—scientists, entrepreneurs, policymakers, and thought leaders—are not just advancing quantum computing; they are redefining the frameworks within which we understand and utilize this transformative technology. This chapter profiles some of the key figures in the quantum realm, highlighting their contributions and visions for the future.

1. Pioneering Scientists and Researchers

The foundation of the quantum revolution rests on the shoulders of pioneering scientists who have dedicated their careers to understanding and harnessing the principles of quantum mechanics.

John Preskill: Often regarded as one of the leading voices in the field, John Preskill, a professor of theoretical physics at Caltech, coined the term "quantum supremacy" and has played a crucial role in establishing the theoretical framework for quantum computing. His work emphasizes the importance of understanding quantum mechanics not only as a scientific discipline but also as a practical tool for solving real-world problems. Preskill envisions a future where quantum computers can tackle challenges beyond the reach of classical computing, such as complex simulations and optimization problems.

Alice Silverberg: A trailblazer in quantum cryptography, Alice Silverberg has contributed significantly to the development of secure communication protocols that leverage the principles of quantum mechanics. Her research focuses on quantum key distribution (QKD), which promises to revolutionize data security. Silverberg advocates for the integration of quantum cryptography into everyday technologies to enhance cybersecurity on a global scale.

Dario Gil: As the Chief Science Officer at IBM, Dario Gil has been instrumental in leading the company's quantum computing initiatives. Under his leadership, IBM has developed a comprehensive quantum ecosystem, including the IBM Quantum Experience, which allows researchers and developers to experiment with quantum algorithms on real quantum hardware. Gil envisions a future where quantum computing is democratized, making it accessible to a broader audience, including businesses and educational institutions.

2. Visionary Entrepreneurs

In addition to scientists, visionary entrepreneurs are crucial in translating quantum research into practical applications and commercial products.

IonQ Founders (Peter Chapman and Chris Monroe): Co-founders of IonQ, a leading quantum computing company, Peter Chapman and Chris Monroe are at the forefront of developing trapped-ion quantum computers. Their vision is to create quantum systems that can outperform classical computers for specific tasks, such as complex data analysis and optimization. They are committed to making quantum computing available as a cloud service, empowering businesses across various sectors to harness its potential.

Michelle Simmons: A pioneer in the development of silicon-based quantum computers, Michelle Simmons leads a research team at the University of New South Wales. Her work focuses on fabricating quantum bits (qubits) using silicon, a material compatible with existing semiconductor technology. Simmons envisions a future where quantum computing is seamlessly integrated into existing technologies, paving the way for advanced applications in fields like artificial intelligence and big data.

Rami El-Etr: As the CEO of Q-CTRL, Rami El-Etr is committed to addressing one of the most significant challenges in quantum computing: qubit stability and error correction. Q-CTRL develops software solutions that optimize quantum operations, making quantum systems more robust and reliable. El-Etr's vision includes creating a future where quantum computers are as reliable as classical systems, enabling widespread adoption across industries.

3. Influential Policymakers and Advocates

The role of policymakers and advocates is crucial in shaping the regulatory landscape and fostering an environment conducive to quantum innovation.

Erik H. Cohen: As a key figure in the U.S. government's quantum initiative, Erik H. Cohen advocates for policies that promote quantum research and development. His efforts focus on securing funding for quantum projects, fostering collaboration between academia and industry, and addressing national security concerns related to quantum technologies. Cohen envisions a future where the U.S. leads in quantum research while ensuring responsible use of the technology.

Luciana B. Florez: An advocate for equitable access to quantum education and research, Luciana B. Florez has dedicated her career to promoting diversity in STEM fields, particularly in quantum computing. She emphasizes the importance of including underrepresented voices in the quantum conversation and believes that a diverse workforce will drive innovation and creativity in the field. Florez's vision includes creating educational programs that empower the next generation of quantum leaders from all backgrounds.

4. Interdisciplinary Collaborators

The quantum revolution is inherently interdisciplinary, requiring collaboration among experts in various fields, including physics, computer science, engineering, and social sciences.

Arvind Krishna: As the CEO of IBM, Arvind Krishna leads initiatives that integrate quantum computing with classical systems. His approach emphasizes the importance of hybrid solutions that combine quantum and classical computing capabilities to solve complex problems in industries such as finance, healthcare, and logistics. Krishna envisions a future where quantum computing becomes an integral part of the technological ecosystem, enhancing existing systems and driving innovation.

Hannah Arendt: A philosopher and ethics expert, Hannah Arendt explores the ethical implications of quantum technologies. She advocates for a multidisciplinary approach to quantum governance, emphasizing the need for ethical frameworks that guide research and applications. Arendt's vision includes fostering dialogues between scientists, ethicists, and policymakers to ensure that quantum technologies are developed responsibly and equitably.

5. The Quantum Visionaries of Tomorrow

The leaders shaping the quantum revolution are not just focused on immediate advancements; they are also laying the groundwork for a future that embraces innovation, ethics, and inclusivity.

Community Building: Many of these leaders emphasize the importance of community building within the quantum ecosystem. They advocate for platforms that encourage collaboration among researchers, students, and industry professionals, fostering an environment of shared knowledge and collective progress.

Global Collaboration: The future of quantum technologies hinges on international collaboration. Many thought leaders stress the need for global partnerships that transcend borders, enabling researchers and companies to share insights, resources, and best practices. Such collaboration will be crucial for addressing the global challenges posed by quantum technologies.

Ethical Frameworks: As quantum technologies continue to advance, ethical considerations must remain at the forefront. Leaders in the field are advocating for the development of ethical frameworks that guide research and applications, ensuring that quantum advancements are aligned with societal values and benefit humanity as a whole.

CHAPTER 46:

Quantum and the Environment: Solving Global Crises

In an era marked by climate change, resource scarcity, and environmental degradation, the need for innovative solutions has never been more urgent. Traditional methods of addressing environmental issues often fall short, hampered by limitations in data processing and simulation capabilities. Enter quantum computing—a technology poised to revolutionize our approach to environmental challenges. This chapter explores how quantum computing can help tackle critical environmental issues, from managing natural resources to combating pollution and fostering sustainable practices.

1. Understanding Environmental Complexities through Quantum Simulations

Environmental systems are inherently complex, involving numerous variables and interactions that can be difficult to model accurately. Quantum computing offers the ability to perform simulations that surpass the capabilities of classical computers, enabling researchers to explore environmental phenomena in greater detail.

Modeling Climate Change: Quantum computers can model complex climate systems with unprecedented accuracy, accounting for intricate interactions between the atmosphere, oceans, and land surfaces. For example, quantum simulations can help scientists understand feedback loops, such as how rising temperatures impact cloud formation and, in turn, affect global temperatures. By providing more accurate climate models, quantum computing can inform better policy decisions and climate mitigation strategies.

Optimizing Resource Management: The management of natural resources, such as water and energy, is crucial for sustainability. Quantum computing can optimize resource allocation by analyzing vast datasets and simulating various scenarios. For instance, quantum algorithms could improve the efficiency of irrigation systems in agriculture by predicting water needs based on real-time environmental data. Such optimization not only conserves resources but also enhances agricultural productivity.

2. Fighting Pollution with Quantum-Enabled Solutions

Pollution is one of the most pressing environmental issues facing the planet today. Quantum computing can play a significant role in combating various forms of pollution by facilitating advanced monitoring, detection, and remediation strategies.

Enhanced Pollution Detection: Quantum sensors are capable of detecting pollutants at extremely low concentrations, far surpassing the sensitivity of classical sensors. For example, quantum-enhanced detection methods can identify trace amounts of harmful substances in air and water, allowing for timely interventions to mitigate pollution. These sensors can be deployed in urban areas to monitor air quality continuously, helping communities respond swiftly to pollution spikes.

Quantum Chemistry for Clean Solutions: Quantum computing's ability to simulate molecular interactions can lead to the discovery of new materials and chemicals that are less harmful to the environment. For instance, researchers can design catalysts that enhance the efficiency of chemical reactions involved in carbon capture and conversion, making these processes more viable for large-scale implementation. By developing greener technologies, quantum computing can contribute to reducing industrial emissions and minimizing environmental impact.

3. Addressing Energy Challenges with Quantum Innovations

The transition to sustainable energy sources is a critical aspect of mitigating climate change. Quantum computing can accelerate the development of new energy technologies and optimize existing systems.

Quantum Optimization for Renewable Energy: Quantum algorithms can optimize energy grids, balancing supply and demand in real time. For example, during peak energy usage, quantum computing can analyze data from multiple sources—such as solar panels, wind turbines, and battery storage—to determine the most efficient distribution of energy. This optimization minimizes waste and enhances the reliability of renewable energy systems.

Advancing Energy Storage Solutions: Quantum simulations can lead to breakthroughs in energy storage technologies, such as batteries and supercapacitors. By modeling the behavior of materials at the atomic level, researchers can identify new materials that enhance energy storage capacity and charge/discharge efficiency. These advancements will be critical for enabling widespread adoption of renewable energy sources, such as solar and wind.

4. Sustainable Agriculture and Quantum Farming

Agriculture is both a significant contributor to environmental issues and a sector heavily impacted by climate change. Quantum computing can help promote sustainable agricultural practices that minimize environmental harm.

Precision Agriculture: Quantum computing can enhance precision agriculture by analyzing vast amounts of data from sensors, satellites, and drones. This analysis can lead to optimized planting schedules, targeted pest control, and precise fertilizer application, reducing waste and environmental impact. For instance, quantum algorithms can determine the ideal conditions for crop growth, allowing farmers to maximize yields while minimizing resource use.

Genetic Research for Resilient Crops: Quantum computing can expedite genetic research by simulating the interactions of genes and their effects on crop traits. This capability can lead to the development of crops that are more resilient to climate change, pests, and diseases. By improving crop resilience, quantum technologies can contribute to food security in the face of changing environmental conditions.

5. Fostering Global Collaboration for Environmental Solutions

The global nature of environmental issues necessitates collaboration among researchers, governments, and organizations. Quantum computing can facilitate this collaboration by providing a platform for shared knowledge and resources.

Quantum Data Sharing Initiatives: Quantum technologies can enable secure and efficient data sharing among researchers and policymakers. This data sharing is essential for understanding regional environmental challenges and implementing effective solutions. By creating a quantum-enabled platform for environmental data, stakeholders can collaborate more effectively to address global issues.

Joint Research Programs: Governments and institutions can establish joint research programs focused on using quantum computing to solve environmental problems. These programs can leverage the expertise of researchers worldwide, fostering innovation and accelerating the development of practical solutions. International cooperation is vital for tackling challenges that transcend national borders, such as climate change and biodiversity loss.

6. Ethical Considerations and Future Directions

While the potential of quantum computing to address environmental issues is vast, ethical considerations must guide its development and implementation.

Equitable Access to Technology: Ensuring that the benefits of quantum technologies are accessible to all, particularly in developing regions, is crucial. Efforts should be made to bridge the technology gap and provide resources for those who may lack access to advanced computing capabilities.

Sustainable Practices in Quantum Development: As the quantum computing industry grows, it is essential to prioritize sustainability in its development. This includes minimizing energy consumption in quantum data centers and considering the environmental impact of manufacturing quantum hardware.

CHAPTER 47:

Quantum Collaboration: Building a Better World

In an interconnected world, the challenges we face—from climate change and public health crises to technological inequities—demand solutions that transcend national and disciplinary boundaries. As quantum computing emerges as a transformative technology with the potential to address some of society's most pressing issues, fostering collaboration across various sectors becomes imperative. This chapter delves into the importance of global cooperation and interdisciplinary collaboration in advancing quantum technologies for the collective good.

1. The Quantum Landscape: A Multidisciplinary Frontier

Quantum computing stands at the intersection of multiple fields, including physics, computer science, engineering, materials science, and even social sciences. Each discipline offers unique insights and methodologies that can contribute to the advancement of quantum technologies.

Cross-Disciplinary Innovation: The most groundbreaking innovations often arise from the convergence of diverse fields. For instance, combining quantum mechanics with machine learning could yield new algorithms that enhance the capabilities of quantum computers. This interdisciplinary approach requires collaboration between quantum physicists and computer scientists, as well as input from experts in ethics and social implications.

Bridging Knowledge Gaps: Quantum computing is still a nascent field, and many organizations may lack the expertise to harness its full potential. Collaborations between academic institutions, industry leaders, and government agencies can bridge these knowledge gaps. By sharing resources, knowledge, and best practices, stakeholders can accelerate advancements in quantum research and its applications.

2. Global Cooperation: Uniting for a Common Purpose

As quantum technologies evolve, global cooperation becomes crucial in addressing shared challenges and ensuring equitable access to quantum advancements.

International Research Initiatives: Collaborative research initiatives can pool resources and talent from around the world. For example, the European Quantum Flagship program aims to foster collaboration among researchers and industries across Europe to advance quantum technologies. Similar initiatives can be established globally, focusing on specific challenges such as climate change, health crises, or cybersecurity threats.

Shared Infrastructure and Resources: Developing a shared quantum infrastructure, such as cloud-based quantum computing platforms, can democratize access to quantum resources. By enabling researchers and organizations from diverse backgrounds to access quantum computing power, we can accelerate innovation and ensure that no one is left behind in the quantum revolution.

3. Quantum Diplomacy: Navigating Global Challenges

As quantum computing technology becomes increasingly integral to national security, economic stability, and technological leadership, it is essential to engage in diplomatic efforts to navigate these challenges.

Establishing International Standards: Countries must work together to develop international standards and protocols for quantum technologies. This includes establishing guidelines for ethical practices, cybersecurity, and intellectual property rights. Collaborative efforts can help mitigate risks associated with the misuse of quantum technologies, such as quantum hacking or the weaponization of quantum systems.

Fostering Global Dialogue: Creating platforms for dialogue among policymakers, industry leaders, and researchers is essential for building consensus on how to approach the societal implications of quantum computing. Conferences, workshops, and collaborative forums can facilitate knowledge exchange and encourage the sharing of best practices for responsible quantum innovation.

4. Engaging Communities: The Role of Education and Awareness

For quantum technologies to benefit society as a whole, it is essential to engage communities and promote education about quantum computing.

Public Awareness Campaigns: Raising awareness about the potential of quantum computing and its implications for various sectors is vital. Public engagement initiatives can demystify quantum technologies, encouraging interest in STEM fields and fostering a new generation of quantum thinkers.

Community-Driven Projects: Involving communities in quantum projects can help ensure that the benefits of quantum advancements are distributed equitably. For example, local universities and research institutions can collaborate with community organizations to develop programs that address specific local challenges using quantum computing.

5. Collaborative Approaches in Addressing Global Issues

Several global issues, such as healthcare, climate change, and poverty, can benefit from collaborative quantum initiatives.

Healthcare Innovations: Collaborative projects leveraging quantum computing can lead to breakthroughs in personalized medicine, drug discovery, and disease modeling. By pooling data from various healthcare institutions, researchers can develop more effective treatment strategies and improve health outcomes on a global scale.

Sustainable Development Goals (SDGs): Quantum technologies can play a vital role in achieving the United Nations' Sustainable Development Goals. Collaborative efforts can harness quantum computing to optimize resource use, enhance energy efficiency, and improve food security, ultimately contributing to global sustainability.

6. The Importance of Ethical Collaboration

As quantum technologies advance, ethical considerations must be at the forefront of collaborative efforts.

Developing Ethical Frameworks: Collaborative initiatives should prioritize the development of ethical frameworks that guide the use of quantum technologies. Engaging ethicists, social scientists, and community representatives can ensure that diverse perspectives are considered in shaping the ethical landscape of quantum innovation.

Addressing Inequities: It is crucial to address potential inequities that may arise from quantum advancements. Collaborative efforts must focus on ensuring that all communities, particularly marginalized ones, have access to the benefits of quantum computing. This includes addressing the "quantum divide" and fostering inclusivity in quantum research and applications.

CHAPTER 48:

The Quantum Society: A New Social Contract

As quantum computing integrates deeper into the fabric of our daily lives, it brings not just technological advancements but a profound transformation in how we think about societal structures, governance, and personal relationships. This chapter envisions a new social contract—a dynamic agreement that evolves alongside technological progress, shaping the interactions between individuals, governments, and corporations in the quantum age.

1. The Shifting Paradigms of Society

The traditional social contract, rooted in Enlightenment thinking, has often centered around the balance of rights and responsibilities between the governed and the governing. In a world increasingly influenced by quantum technologies, this contract must adapt to new realities.

Empowerment through Technology: Quantum computing promises unprecedented computational power and access to information, empowering individuals in ways previously unimaginable. The new social contract must reflect this empowerment, ensuring that citizens can leverage quantum technologies for personal and collective benefits.

Transparency and Trust: With the rise of quantum-enabled data processing, the importance of transparency in government and corporate actions becomes paramount. Citizens will demand clarity regarding how their data is used and the ethical implications of quantum advancements. A new social contract should prioritize transparency, fostering trust between individuals and institutions.

2. Redefining Governance in the Quantum Era

Governments play a crucial role in navigating the complexities introduced by quantum technologies. As quantum computing reshapes industries, governance models must evolve to address these changes effectively.

Agile Policymaking: The rapid pace of quantum innovation requires governments to adopt agile policymaking approaches. Traditional legislative processes may struggle to keep up with the evolving landscape of technology. Emphasizing adaptability and responsiveness will be key to effectively regulating quantum applications while encouraging innovation.

Collaborative Governance: The new social contract calls for collaborative governance models that involve multiple stakeholders, including citizens, tech companies, and academia. Engaging diverse perspectives in policymaking can lead to more holistic solutions that reflect the needs and values of society as a whole.

3. Corporate Responsibility in a Quantum World

As corporations harness quantum computing for competitive advantage, their role in society must also be reexamined. The relationship between businesses, consumers, and the broader community will be influenced by the ethical implications of quantum technologies.

Ethical Innovation: Companies will be held accountable not only for their profits but also for the ethical implications of their innovations. The new social contract must establish frameworks for responsible corporate behavior, ensuring that businesses prioritize ethical considerations alongside technological advancements.

Corporate Citizenship: Businesses must embrace a sense of corporate citizenship, recognizing their responsibilities to the communities they serve. This includes investing in local initiatives, supporting education and workforce development, and addressing societal challenges through innovative solutions made possible by quantum technologies.

4. Data Ownership and Privacy Rights

Quantum computing will significantly impact data processing capabilities, raising critical questions about data ownership and privacy rights in the quantum age.

Personal Data as Property: The new social contract may redefine the concept of personal data ownership. Individuals could be empowered to treat their data as a form of property, with rights to control, monetize, or share it according to their preferences. This shift would necessitate legal frameworks that recognize individual ownership of data and the right to privacy.

Enhanced Privacy Protections: As quantum technologies enable new forms of data analysis, individuals will demand enhanced privacy protections. The social contract must establish clear guidelines for data use, ensuring that individuals' privacy is respected in the face of powerful quantum capabilities.

5. Education and Workforce Development

To thrive in a quantum society, education systems must adapt to prepare future generations for the challenges and opportunities presented by quantum technologies.

Quantum Literacy: The new social contract should prioritize quantum literacy as a fundamental skill, comparable to digital literacy in the current landscape. Educational institutions must integrate quantum concepts into curricula, equipping students with the knowledge and skills needed to navigate the quantum era.

Reskilling and Upskilling Initiatives: As quantum technologies reshape industries, workforce development programs must focus on reskilling and upskilling initiatives. Collaborations between educational institutions, governments, and businesses can create pathways for individuals to transition into emerging roles in the quantum economy.

6. Fostering Social Equity in the Quantum Era

As quantum computing offers transformative possibilities, it is essential to address potential disparities that may arise in access to technology and resources.

Bridging the Digital Divide: The new social contract should actively seek to bridge the digital divide, ensuring equitable access to quantum resources and education across all communities. Initiatives that promote inclusivity and accessibility will help mitigate disparities and ensure that everyone can benefit from the quantum revolution.

Empowerment through Participation: The social contract must encourage active participation from marginalized communities in shaping the future of quantum technologies. By fostering inclusive dialogue and empowering diverse voices, society can create a more equitable and just quantum landscape.

7. The Role of Ethics and Philosophical Considerations

The integration of quantum computing into society raises ethical questions that challenge our understanding of humanity, knowledge, and existence itself.

Navigating Ethical Dilemmas: As quantum technologies open new frontiers, ethical dilemmas will arise regarding the implications of their use. The social contract should incorporate mechanisms for ethical deliberation, allowing society to collectively navigate the complexities of quantum innovation.

Philosophical Reflections: The new social contract may also prompt philosophical reflections on the nature of reality, consciousness, and human identity in a quantum world. Engaging in interdisciplinary discussions that bridge science, philosophy, and ethics will enrich our understanding of the implications of quantum technologies.

CHAPTER 49:

Quantum Futures: Utopias and Dystopias

As quantum computing advances, its transformative potential will shape the fabric of society in unprecedented ways. In this chapter, we will explore two contrasting visions of the future: a utopian society enhanced by quantum technology, characterized by unprecedented prosperity, equity, and interconnectedness, versus a dystopian scenario where the same technology exacerbates social divides, creating a world rife with inequality, surveillance, and fear. By examining these potential futures, we can better understand the choices we face today and the importance of shaping a positive trajectory for quantum technology.

1. Quantum Utopia: A Vision of Abundance and Equality

In this utopian future, quantum technology serves as a catalyst for solving some of humanity's most pressing challenges. Here, we envision a society where the fruits of quantum advancements are distributed equitably, fostering a thriving global community.

Universal Access to Resources: In a quantum utopia, access to computing power and data is democratized. Quantum resources are made available to everyone, regardless of socioeconomic status. Educational institutions leverage quantum computing to provide personalized learning experiences, empowering individuals to develop skills that match the demands of the quantum economy. This emphasis on education reduces skill gaps and creates a highly knowledgeable workforce.

Breakthroughs in Medicine and Health: Quantum computing revolutionizes healthcare, leading to breakthroughs in personalized medicine, rapid drug discovery, and disease prevention. With the ability to model complex biological systems at unprecedented scales, researchers can create tailored treatment plans for individuals, improving overall public health. The reduced cost and increased efficiency of healthcare systems foster well-being for all, eradicating disparities in medical access.

Sustainable Solutions to Climate Change: Quantum technology plays a crucial role in addressing climate change, enabling more efficient energy systems, optimized resource management, and advanced climate modeling. With the help of quantum simulations, we develop sustainable practices that minimize environmental impact and promote ecological balance. Governments and corporations collaborate to implement quantum-driven solutions, creating a greener planet for future generations.

Enhanced Communication and Collaboration: The emergence of a quantum internet facilitates secure and instantaneous communication worldwide. This new network fosters global collaboration on research, innovation, and cultural exchange. As barriers dissolve, communities unite to tackle challenges collectively, resulting in a sense of global citizenship and shared purpose.

Ethical Governance and Social Equity: A new social contract emerges, emphasizing ethical governance and social equity. Policymakers prioritize inclusivity and transparency, involving diverse stakeholders in decision-making processes. Citizens actively engage in shaping their communities, ensuring that technological advancements benefit all rather than a privileged few.

2. Quantum Dystopia: A Future of Surveillance and Division

Contrastingly, we must also confront the potential dystopian consequences of quantum technology. In this scenario, the same innovations that promise progress could deepen existing social divides, leading to a fragmented and oppressive society.

Concentration of Power: In a dystopian quantum future, access to advanced quantum computing becomes the domain of a select few —wealthy corporations and powerful governments. The resulting concentration of power enables surveillance technologies that monitor individuals at an unprecedented scale. Citizens live under constant scrutiny, with their actions tracked and analyzed in real-time, leading to a loss of privacy and autonomy.

Deepening Inequality: The gap between the privileged and the marginalized widens as access to quantum resources remains limited. Those who can afford quantum technology benefit from enhanced capabilities, while the disenfranchised struggle to compete. Education systems fail to adapt, leaving large segments of the population ill-equipped for the quantum economy, perpetuating cycles of poverty and disenfranchisement.

Weaponization of Technology: As quantum computing becomes integrated into military and defense strategies, the risk of arms races escalates. Nations seek to leverage quantum advancements for offensive capabilities, resulting in a new era of warfare characterized by cyberattacks, misinformation, and geopolitical instability. The very technologies designed to foster peace and collaboration become tools of conflict and division.

Loss of Autonomy and Control: The implications of quantum algorithms in decision-making processes raise ethical dilemmas. Automated systems, powered by quantum computing, may make choices on behalf of individuals, diminishing personal agency. People find themselves trapped in systems that prioritize efficiency over human values, leading to feelings of alienation and despair.

Societal Disillusionment: As disparities and injustices persist, societal trust in institutions erodes. Disillusionment with governance, corporations, and technological promises grows, leading to unrest and polarization. Communities fracture along lines of privilege and access, as those left behind feel marginalized and unheard.

3. Choosing Our Future: The Role of Consciousness and Intent

The futures outlined above serve as stark reminders of the power of technology to shape our lives. Whether we arrive at a quantum utopia or a dystopia hinges on the choices we make today.

Collective Responsibility: As stewards of technology, we bear the responsibility to shape its trajectory. This requires collective awareness of the potential consequences of our actions and an active commitment to ethical considerations in the development and deployment of quantum technologies.

Inclusive Dialogue: Engaging in inclusive dialogue across disciplines, cultures, and communities will be essential. By fostering open conversations about the implications of quantum computing, we can navigate the complexities of technological advancement while centering human values and dignity.

Proactive Governance: Policymakers must take proactive measures to address the risks associated with quantum technology, establishing regulations that promote equitable access and protect individual rights. The ethical governance of quantum innovations will require collaboration among governments, industry leaders, and civil society to ensure responsible practices.

Education and Empowerment: Prioritizing education and empowerment is critical for preparing society to navigate the quantum era. By investing in education and skills development, we can equip individuals with the knowledge and tools necessary to thrive in a rapidly changing landscape.

CHAPTER 50:

Embracing the Quantum Age

As we conclude our exploration of quantum computing and its profound implications for society, it is crucial to synthesize the insights we have gained and to reflect on the responsibilities that accompany this revolutionary technology. The quantum age beckons us with immense opportunities and formidable challenges, inviting us to navigate the complexities of a rapidly evolving world with wisdom, hope, and a sense of purpose.

1. A Quantum Perspective: Unleashing Potential

Throughout this book, we have witnessed how quantum computing has the potential to reshape virtually every facet of our lives—from healthcare and climate science to finance, communication, and beyond. The core tenets of quantum mechanics—superposition and entanglement—serve as powerful metaphors for the interconnectedness of our global society. Just as quantum particles exist in multiple states simultaneously, we too must learn to embrace complexity, adaptability, and collaboration in addressing the multifaceted challenges of our time.

Harnessing Innovation: Quantum technology presents us with tools to solve problems previously deemed intractable. By embracing innovation and fostering interdisciplinary collaboration, we can unlock breakthroughs in fields such as medicine, energy, and materials science, enhancing the quality of life for billions of people.

Expanding Horizons: The quantum age expands our horizons, urging us to rethink traditional paradigms and explore new possibilities. As we confront pressing global issues—climate change, health crises, and social inequality—quantum solutions can guide us toward more effective and sustainable approaches, encouraging us to imagine a future where technology uplifts humanity as a whole.

2. A Call to Ethical Responsibility

While the promise of quantum technology is tantalizing, we must approach it with a keen awareness of the ethical implications it entails. The profound capabilities of quantum computing can easily be misused, leading to surveillance, inequality, and ethical dilemmas. As we step into this new era, we must prioritize the following principles:

Ethics and Governance: Establishing robust ethical frameworks and governance structures is imperative to ensure that quantum advancements are deployed responsibly. Policymakers, technologists, and citizens alike must engage in ongoing dialogue to develop regulations that safeguard privacy, prevent abuse, and promote equitable access.

Empowerment Through Education: Education will play a pivotal role in preparing society for the quantum age. By fostering quantum literacy across diverse populations, we can empower individuals to engage with technology thoughtfully and critically. This proactive approach will equip people with the skills necessary to navigate the complexities of the quantum landscape, ensuring that no one is left behind.

Collective Agency: The journey into the quantum era will require collective agency and shared responsibility. We must foster a culture of collaboration where diverse voices are heard, ensuring that technological advancements reflect the values and aspirations of the broader community.

3. The Vision of a Quantum Future

As we embrace the quantum age, we stand at a crossroads, with two divergent paths before us: one leading to a world enriched by quantum possibilities, and the other toward division and conflict. The choices we make today will shape our collective destiny, and we have the power to steer toward a brighter future.

Utopian Possibilities: By harnessing the transformative potential of quantum computing, we can create a society characterized by abundance, equity, and sustainability. Imagine a world where healthcare is personalized, education is accessible to all, and climate solutions are grounded in scientific rigor. This vision is achievable if we work collaboratively and prioritize the common good.

Preventing Dystopian Outcomes: Conversely, neglecting ethical considerations and allowing power imbalances to persist could lead us down a dystopian path. We must remain vigilant against the forces that seek to exploit technology for nefarious purposes, ensuring that our progress does not come at the expense of human dignity or freedom.

4. Moving Forward with Hope

As we move forward into the quantum age, let us do so with a spirit of hope, responsibility, and resilience. The challenges we face are formidable, but they are not insurmountable. Together, we can cultivate a culture of innovation, ethical stewardship, and social equity, embracing the quantum revolution as an opportunity to uplift humanity.

Adaptation and Resilience: The ability to adapt and respond to unforeseen challenges will be paramount in the quantum era. By fostering resilience in individuals and communities, we can navigate uncertainty with confidence, recognizing that collaboration and empathy will be our strongest assets.

Fostering Global Cooperation: In our interconnected world, the quest for a better future transcends borders. Global cooperation will be essential as we tackle shared challenges and collectively envision a society that respects and uplifts all its members. By working together, we can ensure that the benefits of quantum technology are shared equitably across the globe.

Embracing the Quantum Era

As we draw this exploration of quantum computing to a close, we find ourselves at a pivotal moment in human history. The profound advancements in quantum technology not only promise to reshape our understanding of computation but also hold the potential to redefine our societies, economies, and environments. The quantum age is upon us, and with it comes a dual responsibility: to harness its immense power wisely and to navigate the ethical complexities that accompany such transformation.

A New Paradigm of Possibility

The chapters of this book have illuminated the myriad ways in which quantum computing will impact diverse fields—from healthcare to finance, climate science to entertainment. Each innovation offers a glimpse into a future where problems that once seemed insurmountable may become solvable. Quantum computing encourages us to think beyond classical constraints, inviting us to explore solutions that leverage the unique principles of quantum mechanics, such as superposition and entanglement.

As we move into this new paradigm, we must embrace the infinite possibilities that quantum technologies present. By fostering interdisciplinary collaboration and cultivating a culture of innovation, we can unlock breakthroughs that enhance quality of life and address some of the most pressing challenges of our time.

Navigating Ethical Waters

However, the promise of quantum computing is not without its pitfalls. The ethical considerations surrounding privacy, inequality, and security are paramount. As quantum technologies become more integrated into the fabric of society, we must establish robust frameworks that ensure they are developed and implemented responsibly. It is essential that we address the risks of a potential quantum divide, where access to these technologies is restricted to a privileged few, exacerbating existing social inequalities.

We must prioritize education and awareness, empowering individuals with the knowledge and skills necessary to engage thoughtfully with quantum technologies. By cultivating a collective understanding of the ethical implications and potential consequences of our choices, we can navigate this complex landscape with a commitment to justice, equity, and human dignity.

A Vision for Global Cooperation

The journey into the quantum era calls for global cooperation. As we face shared challenges, such as climate change and public health crises, the potential for quantum computing to foster solutions necessitates collaboration across borders and disciplines. Institutions, governments, and individuals must come together to harness the benefits of quantum technology for the collective good, ensuring that no one is left behind in the pursuit of progress.

As we envision a future shaped by quantum advancements, let us commit to a path of inclusivity and cooperation, where the fruits of innovation are shared equitably and contribute to the well-being of all.

A Collective Responsibility

Ultimately, the quantum age offers us an opportunity to redefine our relationship with technology and with one another. It challenges us to reflect on our values and to cultivate a future where technological advancement is aligned with our deepest ethical commitments.

As we stand on the threshold of this new era, let us approach it with curiosity, responsibility, and hope. The journey ahead may be complex and fraught with uncertainty, but it is also filled with promise. By embracing the transformative potential of quantum computing while remaining vigilant in our ethical considerations, we can forge a future that enhances human flourishing and fosters a sustainable, equitable, and compassionate world.

In conclusion, let us step boldly into the quantum age, ready to shape our destinies and to create a legacy of innovation and integrity for generations to come. The future is quantum; let us embrace it together.

ABOUT THE AUTHOR

Laura Lee was born in a small town, where her early life was shaped by a close-knit community and a strong sense of local values. Growing up in an environment that valued hard work and personal connection, Laura developed a deep appreciation for community and collaboration. With a passion for storytelling and a keen interest in human behavior, she pursued a career in writing and motivational speaking. Her work reflects her commitment to personal growth and empowerment, drawing from her experiences and insights gained from her humble beginnings. Laura's unique perspective and heartfelt approach have made her a respected voice in her field, inspiring others to embrace their potential and pursue their dreams.